Created and Natural Wetlands for Controlling Nonpoint Source Pollution

■

Edited by
Richard K. Olson

U.S. EPA
Office of Research and Development
and
Office of Wetlands, Oceans, and Watersheds

C. K. SMOLEY

Library of Congress Cataloging-in-Publication Data

Created and natural wetlands for controlling nonpoint source pollution / Office of Wetlands, Oceans, and Watersheds. U.S. Environmental Protection Agency.
p. cm.
Includes bibliographical references.
ISBN 0-87371-943-3 (alk. paper)
1. Water quality management—United States. 2. Water—Pollution—United States. 3. Wetland conservation—United States. 4. Constructed wetlands—United States. I. United States. Environmental Protection Agency. Office of Wetlands, Oceans, and Watersheds.
TD223.C73 1993
628.1'68—dc20 92-31139
CIP

Direct all inquiries to CRC Press, Inc.
2000 Corporate Blvd., N.W.
Boca Raton, FL 33431

PRINTED IN THE UNITED STATES OF AMERICA
1 2 3 4 5 6 7 8 9 0
Printed on acid-free paper

Table of Contents

Acknowledgment

This publication was prepared and assembled by Richard K. Olson, ManTech Environmental Technology, Inc., USEPA Environmental Research Laboratory, Corvallis, Oregon, with Technical Editor Kay Marshall of Technical Resources, Inc., Rockville, Maryland.

Notice

This document has been subjected to peer and administrative review and approved for publication by the U.S. Environmental Protection Agency Research Laboratory in Corvallis, Oregon, through ManTech Environmental Technology, Inc. and Technical Resources, Inc. Mention of trade names or commercial products does not constitute endorsement or recommendation for use.

CHAPTER 1

Evaluating the Role of Created and Natural Wetlands in Controlling Nonpoint Source Pollution

Richard K. Olson, ManTech Environmental Technology, Inc.

ABSTRACT

Nonpoint source (NPS) pollution control and wetlands protection are two overlapping scientific and policy issues of the U.S. Environmental Protection Agency. Created, restored, and natural wetlands can contribute significantly to watershed water quality, but at the same time must be protected from degradation by NPS pollution. Effective use of wetlands in NPS control requires an integrated landscape approach including consideration of social, economic, and government policy issues as well as scientific knowledge.

INTRODUCTION

This Proceedings addresses two issues that are major environmental concerns and major policy issues for the U.S. Environmental Protection Agency (EPA):

1. Nonpoint source (NPS) pollution contributes over 65% of the total pollution load to U.S. inland surface waters (US EPA, 1989). Sources include urban stormwater, diffuse agricultural runoff from pastures and

row crops, concentrated agricultural wastes from feedlots, runoff from building sites, forestry activities, and drainage from mining activities.

2. Over half of the wetlands in the lower 48 states have been lost during the past 200 years (Dahl, 1990), with some states losing more than 85 percent of their wetlands. Remaining wetlands are frequently degraded through physical alteration, hydrologic modification, and exposure to pollutants.

Addressing these two issues in an integrated manner makes sense in terms of both science and policy. Wetlands occupy depressions in the landscape and therefore are often recipients of waterborne pollutants. This exposure, coupled with the inherent ability of wetlands to sequester or transform many pollutants, gives them an important role in water quality improvement. This role can be consciously used in strategies to control NPS pollution by creating, restoring, or preserving wetlands in appropriate locations in the landscape. The other side of this issue is that wetlands can be degraded by NPS pollution. Preventing degradation of the full range of wetland functions may be at odds with maximizing their water quality functions.

The policy linkages result from Sections of the Clean Water Act (CWA) (33 U.S.C. 1251 as amended) that give EPA responsibilities in wetland protection and NPS pollution control (US EPA, 1990). Under the CWA, most wetlands are considered to be "waters of the U.S." (Bastian et al., 1989) and are included within the CWA's objective to "restore and maintain the chemical, physical, and biological integrity of the Nation's waters." Section 404 of the CWA regulates the discharge of dredge and fill materials, and Sections 401 and 402 the discharge of waterborne pollutants into "waters of the U.S..." Under Section 404, restoration or creation of wetlands may be required to mitigate wetland losses.

Section 319 provides a mechanism for integrating Federal and State programs for controlling NPS pollution. States are required to perform, under EPA oversight and approval, assessments of the status of NPS pollution, and to develop management programs to control NPS pollution. Wetland protection, creation and restoration may be included in State 319 programs. Thus, the 404 program and related efforts may protect and restore wetlands and their water quality functions, to the benefit of NPS control programs. The NPS programs can protect wetlands from degradation by pollutants, while also identifying areas where wetland creation and protection will optimize wetland water quality, functions.

WORKSHOP

In order to further define the scientific and policy linkages between NPS pollution and wetlands issues, the EPA Office of Research and Development and the Office of Wetlands, Oceans and Watersheds held a workshop on 10-11 June, 1991, in Arlington, Virginia. Within the overall theme of evaluating the role of created and natural wetlands in the control of rural NPS pollution, the workshop objectives were:

1. To review the state of knowledge; and
2. To identify research needs and approaches for developing guidelines for the inclusion of wetlands in NPS control strategies.

The focus of the workshop was rural NPS pollution, exclusive of acid mine drainage. Urban stormwater and acid mine drainage are both important pollution sources for which wetlands can provide treatment. However, urban stormwater not only differs somewhat from rural NPS in its chemical constituents, but by definition occurs in a landscape from which many of the options for creation and use of wetlands have been lost. Acid mine drainage is very different chemically from most rural NPS pollution, and its treatment by wetlands requires consideration of a number of different issues.

Restricting the workshop scope in this way still left for consideration a variety of nonpoint sources ranging in scale from local (e.g., swine farms) to watershed and regional, and a large number of scientific and policy questions concerning the use of wetlands to treat NPS pollution. Presentations during the first part of the workshop provided background on these issues, and papers corresponding to each presentation are included in this volume.

WORKSHOP RESULTS

Participants in the workshop included staff from EPA wetlands and NPS programs; representatives of the U.S. Department of Agriculture, U.S. Fish and Wildlife Service, and the U.S. Army Corps of Engineers; and wetlands scientists from universities and consulting firms. This diversity of backgrounds and perspectives led to a corresponding diversity of ideas and opinions. A number of themes emerged, however, as unifying threads throughout the discussions and the papers in this volume. These include:

- Natural wetlands should not be used as wastewater treatment systems. In most cases, natural wetlands are considered "waters of the U.S.," and are entitled under the CWA to protection from degradation by NPS pollution. Natural wetlands do function within the watershed to improve water quality, and protection or restoration of wetlands to maintain or enhance water quality are acceptable practices. However, NPS pollutants should not be intentionally diverted to these wetlands, and wetlands receiving NPS loadings that will degrade the wetland should be protected by establishing upland buffer strips or other best management practices (BMPs).
- Wetlands must be part of an integrated landscape approach to NPS control. Created, restored, and natural wetlands can contribute significantly to watershed water quality, but they must be sited correctly and not be overloaded. Wetlands cannot be expected to compensate for insufficient use of BMPs such as conservation tillage, grassed waterways, and exclusion of livestock from riparian areas.
- The technical and scientific issues involved in defining the role of wetlands in NPS control will be relatively easy to resolve compared to the social and economic issues. Large-scale wetland creation or restoration efforts are expensive. Owners of key restoration sites may be unwilling to participate. Watersheds are a scientifically logical unit for NPS/wetlands programs, but rarely correspond with existing administrative units (e.g., farms or counties).
- Knowledge of technical issues is uneven. Although more research is needed, design criteria for constructed wetland treatment systems have been fairly well worked out. Methods manuals for wetland creation and restoration are being developed by several Federal agencies, and much is known about the fate and effects of nutrients in wetlands.

Topics with less sufficient information include water quality functions at the landscape scale. For example, improved models of sources and movement of NPS pollutants, and the role of wetlands in altering that movement, are needed to guide siting decisions for wetland creation and restoration. The fate and effects of toxics (e.g., pesticides) in wetlands are not well known. Watershed-level demonstrations of the effects of wetland restoration and creation on water quality are lacking, but are needed both as a framework for research programs and as a technology transfer tool.

CONCLUSIONS

The combination of wetlands and NPS pollution makes for complex scientific and policy issues. Technical issues must be addressed within the broader social and economic context if the research results are to contribute to improvements in water quality and wetlands protection. Mechanisms need to be developed to provide a strong link between science and policy development. Successful implementation of NPS control strategies involving wetlands will require participation of citizens, landowners, scientists, and government officials at every stage of the process.

REFERENCES

Bastian, R. K., P. E. Shanaghan, and B. P. Thompson, 1989. Use of wetlands for municipal wastewater treatment and disposal-regulatory issues and EPA policies. pp. 265-278. In: D. A. Hammer (ed)., Constructed Wetlands for Wastewater Treatment: Municipal, Industrial and Agricultural. Lewis Publishers, Inc., Chelsea, MI.

Dahl, T. E., 1990. Wetlands Losses in the United States: 1780's to 1980's. U.S. Department of the Interior, Fish and Wildlife Service, Washington, DC. 21 pp.

U.S. Environmental Protection Agency, 1989. Focus on nonpoint source pollution. The Information Broker, Office of Water Regulations and Standards, Nonpoint Sources Control Branch, Washington, DC., November, 1989.

U.S. Environmental Protection Agency, 1990. National Guidance: Wetlands and Nonpoint Source Control Programs. Office of Water Regulations and Standards, and Office of Wetlands Protection, Washington, DC.

CONCLUSIONS

[illegible]

[illegible]

CHAPTER 2

Introduction to Nonpoint Source Pollution and Wetland Mitigation

Lawrence A. Baker, Water Resources Research Center, University of Minnesota

ABSTRACT

Nonpoint source (NPS) pollution is the major cause of impairment of U.S. surface waters. The dominant source of NPS pollution is agricultural activity, and "traditional" pollutants—nutrients, sediments, and pathogens—are the main detrimental constituents. Erosion from cropland has been declining and is expected to decline further in the 1990s, but it is unclear how this will translate into changes in sediment yields in streams. Pollution by nitrogen is of particular concern in eutrophication of estuaries, as a contaminant of groundwater, and as an acidifying agent in atmospheric deposition. Nitrogen fertilizer and emissions of nitrous oxides are major contributors to the problem. The outlook on pesticides is mixed: bans on organochlorine pesticides in the 1970s have resulted in decreasing concentrations in fish tissue; however, herbicides are now a problem for some surface and groundwater sources of drinking water, especially in the Upper Midwest. Metals in NPS pollution are primarily a concern in mining areas and in urban runoff. Declining use of leaded gasoline has resulted in decreased lead in fish tissues, sediments, and surface waters around the nation. New directions in controlling NPS pollution include: (1) a greater emphasis on risk assessment, (2) a move toward regulatory or quasi-regulatory approaches, and (3) a trend toward source reduction. The potential for using wetlands to control agricultural NPS pollution is discussed by contrasting cropland runoff with secondary wastewater effluent.

INTRODUCTION

During the past 20 years, most of our effort to control water pollution has been directed at reducing point source discharges to surface waters. Today, 144 million people, twice as many as in 1972, are served by municipal wastewater treatment plants that provide treatment at the secondary level or better (U.S. EPA, 1990a). Less than 1 percent of municipal wastewater is now discharged with no treatment. This upgrading of sewage treatment plants, generally from primary to secondary treatment, has resulted in a 46 percent reduction in the discharge of oxygen-consuming pollutants from sewage treatment plants between 1972 and 1982. Had these improvements not occurred, discharges of oxygen-consuming pollutants would have increased by 191 percent due to population growth (ASIWPCA, 1985). But having made progress in this area, we are left with the problem of nonpoint sources of pollution – contaminated runoff from urban areas, agricultural fields, animal feedlots, roadways, abandoned mines, silviculture, and construction activities. Nonpoint source (NPS) pollution has been identified as the major remaining cause of surface water impairment (ASIWPCA, 1985; U.S. EPA, 1989), but resources allocated for control of NPS pollution account for only about 4 percent of our national water pollution control expenses (Farber and Rutledge, 1988; Figure 1).

This paper presents an overview of the status of NPS pollution in the United States and briefly compares the potential of wetlands for removing pollutants from secondary wastewater with their potential for reducing pollutant loadings from cropland. It begins with a review of several national assessments, including recent reports developed by the U.S. Environmental Protection Agency (EPA) and the states as mandated by sections 305(b) and 319 of the Clean Water Act. This is followed by a closer examination of several types of NPS pollution – sediments, nutrients (particularly nitrates), pesticides, salinity, and metals. Where possible, trends for several pollutants are examined to address the question: Is NPS pollution getting better or worse? New directions in the control and monitoring of NPS pollution are discussed briefly. The second part of the paper considers several issues related to the use of constructed or managed wetlands to treat NPS pollution. This section emphasizes some fundamental differences between the use of wetlands for tertiary treatment of municipal wastewater, an area in which we have considerable experience, and their use in mitigating NPS pollution from rural lands, particularly agricultural land, an area in which we have little experience.

NONPOINT SOURCE POLLUTION: STATUS AND TRENDS

About 30 percent of assessed U.S. surface waters do not "fully support" their designated uses (U.S. EPA, 1990a; Table 1). The EPA has concluded that for roughly two-thirds of impaired waters, the cause of impairment is NPS pollution (U.S. EPA, 1986; Figure 2). The EPA's most recent biennial report on NPS pollution (U.S. EPA, 1990b), which is summarized in Figure 3 and Table 2, reports NPS pollution impacts for 206,179 river miles, 5,300,000 acres of lakes, and 5,800 square miles of estuary. One conclusion reached by these reports is that the traditional pollutants, particularly nutrients and sediments, are the primary causes of surface water impairment.

Although the methodologies used in these reports differ considerably, they concur in the finding that agriculture is the largest single cause of use impairment in assessed rivers and lakes (Table 2). A recent National Academy of Sciences report on alternative agriculture (NRC, 1989) emphasizing the negative influence of current national agricultural policies on environmental problems has further heightened interest in agricultural pollution.

Other sources of NPS pollution are more important than agricultural sources in several areas of the country, although they contribute to < 10 percent of the impairment of the assessed national aquatic resources. These sources include urban runoff, land disposal, hydromodification, and mining. A substantial percentage of systems impaired by NPS pollution have an unknown source (Table 2).

The methodologies used in these reports present several problems: (1) only a small portion of the total resource was assessed, (2) there is no consistent methodology for designating water uses (e.g., warmwater fisheries, coldwater fisheries, etc.), (3) there is no uniform methodology for determining use attainment, and (4) because of reporting differences between the 305(b) report (U.S. EPA, 1990a) and the 319 report (U.S. EPA, 1990b), results from the two cannot be directly compared (see U.S. EPA, 1990a). Nevertheless, these reports and the earlier assessment by the ASIWPCA (1985) give a clear impression that NPS pollution, particularly agricultural pollution, is a major source of water quality impairment in the United States.

Several recently completed national studies focus on particular pollutants and resources and give us finer resolution of the scope of the NPS problem. These include two major studies of trends in surface water quality at several hundred stream sites in the U.S. Geological Survey's (USGS) National Stream Water Quality Accounting Network (NASQAN) and National Water Quality Surveillance System (NWQSS), henceforth referred to in aggregate as the USGS network (Smith et al., 1987; Lettenmaier et al., 1991); the National Contaminant Biomonitoring Survey conducted by the U.S. Fish and Wildlife Service (USFWS) (Schmitt et al., 1990; Schmitt and Brumbaugh, 1990); and the EPA's National Groundwater Pesticide Survey (U.S. EPA, 1990c).

Sediments

Sedimentation has been identified as a major source of NPS impairment of U.S. rivers and lakes (Figure 3). Excessive sedimentation results in destruction of fish habitat, decreased recreational use, and loss of water storage capacity. The U.S. Department of Agriculture (USDA) has estimated that annual offside costs of sediment derived from cropland erosion alone are $2-6 billion, with an additional $1 billion arising from loss in compared productivity (USDA, 1987). Erosion from agricultural lands has declined since the dust bowl years, from more than 3.5 billion tons/year in 1938 to 3.0-3.1 billion tons/year in the 1980s (USDA, 1990). Lee reported that cropland erosion declined by 12 percent between 1982 and 1987 (Lee, 1990). This occurred primarily as a result of decreased erosion from land that was continuously cropped throughout the study period. In comparison, conversion of land (cropped land to noncropped land) was relatively unimportant in the overall decline in erosion during 1982-1987. The use of conservation tillage increased during the 1980s, from around 40 million acres in 1980 to 88 million acres in 1988, when it accounted for 30 percent of all cropland. Conservation tillage is expected to increase in the future (USDA, 1990) and when combined with expanded implementation of other conservation compliance programs should result in further reductions in cropland erosion during the 1990s (USDA, 1990).

Does declining erosion from agricultural land result in improved conditions in rivers and lakes? From a water quality standpoint, the relationship between gross erosion (defined as a loss of soil from a parcel of land) and sediment yield (defined as the loss of suspended solids from the watershed) is complex; changes in erosion do not necessarily translate simply into changes in sediment yield. This discrepancy is best illustrated by a classic historical analysis of erosion and sediment yield in the agricultural watershed of Coon Creek, Wisconsin (Trimble, 1981). During the period 1853-1938, annual sheet and rill erosion was 630 x 10^3 tons. This amount declined to 456 x 10^3 tons as a result of improved soil conservation, a decrease of 28 percent (Figure 4). However, sediment yield remained virtually constant (40-42 x 10^3 tons/year) throughout the period of record because stream bank and channel erosion increased when sheet and rill erosion declined (Figure 4). From a broader perspective, it has been estimated that nationally 25 percent of sediment yield occurs from stream bank erosion (Figure 7-10 in Van de Leaden et al., 1990).

Stream modifications, such as channelization and reservoir construction, also affect sediment yield. For example, the completion of reservoirs on the Missouri River during the 1950s and 1960s was probably a major contributing factor in an observed 50 percent decline in sediment discharges by the Mississippi River to the Gulf of Mexico (Made and Parker, 1985). Finally, other land uses account for about half the overail sediment loading to U.S. surface waters, placing an upper limit on the potential for reducing sediment yields by controlling cropland erosion (Van de Leaden et al., 1990). The USDA has estimated that a reduction of one billion tons in gross erosion from the Nation's cropland would cause a 13-18 percent decline in the production of sediments, with concomitant reductions in total phosphorus and total organic nitrogen of 5-7 percent and 7-9 percent, respectively (USDA, 1990).

With this perspective, it is perhaps not surprising that concentrations of suspended solids (SS) have not changed at most USGS network stations since the 1970s (Table 3), despite declining cropland erosion. Smith et al. (1987) and Lettenmaier et al. (1991) both showed that concentrations of suspended solids did not change significantly at most of the USGS network sites. Among remaining sites, roughly the same number have shown an increase in suspended solids as have shown a decrease (Table 3). Smith et al. (1987) observed that increasing SS concentrations tend to occur in areas with high rates of soil erosion.

Nutrients

Phosphorus

Nutrients have been identified as the dominant cause of impairment by NPS pollution in lakes and estuaries (Figure 3). In the majority of freshwater lakes, phosphorus (P) is the limiting nutrient for algal growth (Chiaudani and Vighi, 1974; Miller et al., 1974), an exception being lakes highly enriched in municipal wastewater, which typically have low N:P ratios (Miller et al., 1974; Baker et al., 1985b). There is no national data base on trends in lake eutrophication. In the USGS stream trend studies of Smith et al. (1987) and Lettenmaier et al. (1991), total phosphorus (TP) concentrations paralleled SS concentrations: the majority of stations (75-80 percent) exhibited no change; of those that changed, more decreased than increased (Table 3). However, Smith et al. (1987) concluded that reductions in TP concentrations were probably associated with declines in point sources. For the Great Lakes, municipal point sources of phosphorus were reduced by 51-67 percent between 1975 and 1985 (CEQ, 1990), resulting in major reductions in lake phosphorus levels. Further reductions will depend upon reductions in NPS phosphorus loadings, which in 1986 comprised 59-88 percent of total phosphorus loadings in the Great Lakes (CEQ, 1990).

Where increases in phosphorus occurred, Smith et al. (1987) found statistical associations between TP increases and measures of fertilized acreage and cattle population. Thus, there is some suggestion that phosphorus loadings from agricultural areas are increasing.

Nitrogen

NPS nitrogen pollution is important in at least three arenas: (1) eutrophication of surface waters, (2) groundwater contamination in agricultural areas, and (3) acidification of forested watersheds. Nitrogen is particularly important in regard to the eutrophication of estuaries, since algal growth in estuaries is usually nitrogen limited (reviewed in Stoddard, 1991). An interesting aspect of this problem in the Chesapeake Bay is that atmospheric deposition appears to be a major source of nitrogen (Stoddard, 1991). Direct deposition of nitrogen (NH_4^+ and NH_3^-) to the water surface of the Bay accounts for about 12 percent of the total N loading; additional atmospheric nitrogen loading comes from deposition to the watershed. Although about 90 percent of the nitrogen deposition to the Chesapeake Bay watershed is retained in plants and soils, the remainder passes through the watershed and into the Bay. This accounts for an additional 22 percent of the Bay's nitrogen budget, which together with direct deposition to the Bay surface means that 35 percent of the Bay's total nitrogen loading is derived from atmospheric deposition (Stoddard, 1991; Table 4). Other nonpoint sources (e.g., animal waste, fertilizer fluxes, etc.) account for 16 percent of total nitrogen inputs.

These findings suggest that reductions of nitrous oxide emissions, which are generated primarily by automobiles and other vehicles, would ameliorate the eutrophication problem in the Chesapeake Bay.

One of the greatest NPS pollution concerns in the Midwest and other agricultural areas is nitrate contamination of groundwater. Nielsen and Lee (1987) used USGS well water records in conjunction with an analysis of sensitivity factors to estimate the potential for nitrate contamination in groundwater. In 474 counties out of 1,663 agricultural counties with adequate data, measured nitrate levels were >3 mg/L in more than 25 percent of the wells; 87 of these counties had measured nitrate levels greater than the EPA's maximum contaminant level (MCL) of 10 mg/L in more than 25 percent of the wells. These counties are located primarily in the Great Plains, the Corn Belt, the Southwest, and the Northwest. More recently, the National Pesticide Survey, a statistically designed survey of pesticides and nitrate in drinking water wells of the United States, showed that more than half had detectable levels of nitrate; in 2.4 percent of domestic rural wells (254,000 wells) and 1.2 percent of community supply wells (1,130 wells), nitrate exceeded 10 mg/L (U.S. EPA, 1990c; Table 5).

Finally, elevated atmospheric nitrogen inputs may also result in nitrogen saturation of forested watersheds. Nitrate levels are elevated in atmospheric deposition throughout the eastern United States, occurring in a SO_4^{2-}:NO_3^- ratio of about 2:1 (reviewed in Baker et al., 1991). Nitrate is generally thought to be efficiently retained in watersheds, neutralizing inputs of HNO_3 (Baker et al., 1991). However, recent evidence suggests that high inputs of nitrate may exceed the demand by watershed plants and microbes, resulting in nitrate saturation and subsequent breakthrough of HNO_3 into streams and lakes (Stoddard, 1991). This process is potentially important because the additional inputs of HNO_3 would exacerbate lake and stream acidification. There is conclusive evidence of nitrogen saturation in parts of northern Europe. However, in the United States, where atmospheric nitrate inputs are generally lower, the evidence for nitrogen saturation is sketchy. Some evidence that nitrogen saturation may be occurring has been found in the Catskill and Adirondack Mountains of New York and in the Mid-Appalachian and Smoky Mountains in the Southeast (Stoddard, 1991).

Is nitrogen pollution getting better or worse? Several lines of evidence suggest that it is getting worse, at least for surface waters. First, surface water concentrations increased in the USGS network stations: far more stations exhibited increased nitrate concentrations during 1974-1981 than decreases, and far more stations exhibited increases in total nitrogen during 1978-1984 than declines (Table 3). It is perhaps even more important that rivers discharging into estuaries on the East Coast and the Gulf of Mexico exhibited increases in nitrate concentrations of 20-46 percent (Smith et al., 1987). The probable cause of this increase gayest the increased use of nitrogen fertilizers, from around 6.5 million tons in 1970 to 9-12 million tons in the 1980s (USDA, 1990). Smith et al. (1987) found strong associations between increased nitrate and measures of agricultural activity, supporting the contention that fertilizer nitrogen is a major cause of the uptrend.

Salinity

Excessive salinity is a problem primarily in arid regions, where it can lower crop yields, add to water treatment costs, and increase the maintenance cost of water supply systems. The USGS network studies (Table 2) show that 2-3 times more stations have experienced increases in chloride (a good surrogate for salinity) than have experienced declines. In the western United States, salinity problems are caused largely by irrigated agriculture, which both adds salt load and causes reduction in flow through evaporation. In the Colorado River Basin, damages due to salinity are conservatively estimated at $311 million/year (USDI, 1989). Salinity in the Colorado Basin declined during the early 1980s due to unusually high flows, but salinity levels are expected to increase over the next 20 years as a result of continued development and a return to more normal hydrologic conditions. Salinity control has reduced the salt load by 156,000 tons/year, but an additional million-ton reduction will be needed by the year 2010 to keep salinity at Imperial Dam (near the U.S.-Mexico border) below the criterion level of 879 mg/L.

Increased chloride concentrations in many eastern U.S. streams are probably associated with increased use of road salt (Smith et al., 1987). In several regions of the Equational Stream Survey (a randomized sampling of streams in the eastern United States) 15-30 percent of the lower stream reaches had chloride concentrations more than 10 times higher than would be expected from natural sources (A. T. Herlihy, Utah State University, pers. comm.), indicating widespread chloride contamination in this part of the country.

Pesticides

As with nitrate, there is widespread public concern about contamination of groundwater by pesticides in agricultural areas. Of particular interest in recent years has been the potential for herbicide contamination, because herbicide use has increased fourfold since 1966. However, concerns about increased use of herbicides in conjunction with expanded use of conservation tillage may be overstated, as several studies indicate that total herbicide use may not change appreciably with the acceptance of conservation measures (Fawcett, 1987; Logan, 1987). A compilation of existing data by Williams et al. (1988), summarized in NRC (1989), shows that 46 pesticides have been detected in groundwater from 26 states; atrazine, aldicarb, and alachlor were the most commonly detected. Nielsen and Lee (1987) concluded that the potential for pesticide contamination was greatest in the Eastern Seaboard, the Gulf States, and the Upper Midwest. In the EPA's National Pesticide Survey (U.S. EPA, 1990c), 12 of the 126 pesticides and pesticide metabolites analyzed were found in detectable quantities. One or more of these pesticides were detected in 10.4 percent of the nation's community water supply wells and in 4.2 percent of rural domestic wells. None of the community water supply wells and only 0.6 percent of the rural domestic wells exceeded maximum contaminant levels (MCL) or health advisory limits (HAL) for pesticides. Five pesticides exceeded the MCL/HAL limits in rural domestic wells: atrazine, alachlor, DCPA acid metabolites, lindane, and EDB (Table 5). Thus, from a national perspective, a small number of pesticides has contributed significantly to contamination of groundwater. Two of these pesticides, DBCP and EDB, are no longer in use in the United States.

In the Midwest, surface water concentrations of herbicides commonly exceed proposed standards, at least during part of the year. Baker et al. (1985a) reported seasonally elevated concentrations of atrazine in two Ohio rivers; peak concentrations of several pesticides in finished tap water derived from surface water exceeded what are now proposed MCL/HAL limits. In an ongoing study, the USGS measured herbicide concentrations in 150 rivers in the Midwest, an area that accounts for about 60 percent of the pesticides (mostly herbicides) used in the nation (Goolsby and Thurman, 1990). Concentrations of all measured herbicides were low during the pre-application period, but increased following herbicide application (May-June). During this perjod, nearly half the samples had atrazine concentrations above the HAL limit of 3 μg/L (median = 3.8 μg/L); alachlor, cyanazine, and simazine also exceeded HAL or proposed MCL limits (Figure 5). Although runoff from farmland is undoubtedly the major mode of transport for pesticides, herbicides are also distributed by aerosol drift and by volatilization, which typically accounts for losses of 10-30 percent of field application for many pesticides (see review by Grover, 1991). The USGS recently reported that the average concentration of triazine herbicides in Upper Midwest precipitation was 0.5 μg/L (Capel, 1991).

The situation for many organochlorine compounds whose use was discontinued during the 1970s and early 1980s is encouraging. Results from the National Contaminant Biomonitoring Program (Schmitt et al., 1990), a periodic survey of fish tissue contaminants from 112 stations located throughout the nation, show that concentrations of organochlorine pesticides in fish have generally declined. Among those that declined significantly between 1976 and 1984 were DDT, PCBs, toxaphene, chlordane, and endrin (Figure 6). Schmitt et al. (1990) noted that the presence of p,p'-DDT in some of their fish samples suggested recent inputs. Rappaport et al. (1984) postulated that continuing input of DDT to the United States occurs by atmospheric transport from Central America, where it is still used.

Metals

In the National Water Quality Inventory, the States generally considered metals contamination to be a relatively minor cause of surface water impairment; metals accounted for around 7-8 percent of NPS impairment of rivers and lakes (U.S. EPA, 1990a). However, NPS metal contamination is a significant issue with respect to urban runoff and runoff from mining sites. In a comprehensive study of urban runoff in 22 cities, the EPA concluded that copper, lead, and zinc in urban runoff posed a significant threat to aquatic life. Median concentrations were 34 μg/L, 144 μg/L, and 160 μg/L for copper, lead, and zinc, respectively. Each of these metals exceeded criteria for the protection of aquatic life in more than half the collected runoff samples (U.S. EPA, 1983). Probably of greater significance in terms of the total number of miles of impaired streams and rivers is the leaching of metals from abandoned coal mines in the Appalachian region and abandoned metals mines in the West (Moore and Luoma, 1990). Studies by the U.S. Fish and Wildlife Survey and the Appalachian Regional Commission (reviewed in Herlihy et al., 1990) indicate that there are approximately 10,000 km of acidic mine drainage streams in the Appalachian region. Pollution by metals, particularly selenium, is a problem in some wetlands receiving irrigation return flow, but the problem appears to be limited to closed-basin systems in the West (Deason, 1989).

Undoubtedly, the most important NPS problem of metals has been the atmospheric deposition of lead resulting from combustion of leaded gasoline. Perhaps one of the most successful efforts to control NPS pollution was the removal of lead from gasoline during the early 1970s. The decline in lead use resulted in decreased concentrations in surface waters at the USGS network stations (Smith et al., 1987; Lettenmaier et al., 1991), in Mississippi River delta sediments (Treffry et al., 1985), and in fish tissues in the National Contaminant Biomonitoring Program (Figure 7).

NEW DIRECTIONS

In the nearly 20 years since passage of the Clean Water Act, there has been uneven progress in controlling NPS pollution. However, past experiences have led to a recent refocusing of efforts in several areas, including (1) development of a risk assessment approach to pollution control, (2) a trend towards source reduction as a key to reducing pollution, and (3) a shift from voluntary efforts to regulation and monetary incentives.

Risk Assessment Framework

EPA is gradually developing a risk assessment approach to dealing with pollution problems, so that the effort to control a particular type of pollution will have some relation to the risk posed to humans or ecosystems (SAB, 1990). As a starting point, there is a need to move beyond chemical monitoring to measures of ecological condition (U.S. EPA, 1989; Hughes and Larsen, 1988; Larsen et al., 1988). This is a high priority if we are to determine impacts from diffuse sources, which are often caused by multiple pollutants, commonly in conjunction with habitat alteration. States are currently required to develop biological criteria ("biocriteria") under Sections 303 and 304 of the Clean Water Act, although research to support this effort has been limited (GAO, 1990).

A second component of risk assessment is comparative risk analysis. On a national scale, the 305(b) and 319 Program reports are intended to Provide a comparative risk assessment of water quality problems, although as noted earlier, there are severe deficiencies in the methodologies used in these reports and in most other monitoring programs used for assessment of pollution (Hren et al., 1990). EPA's Ecological Monitoring and Assessment Program (EMAP; U.S. EPA, 1991), now under development, will provide a statistically based representation of ecological conditions in the Nation's waters. The goal of this national-scale sampling effort is to evaluate status and trends in ecological conditions, but EMAP also is designed to address specific issues, such as causes and sources of impairment by NPS pollution, through more detailed diagnostic studies. These studies should be useful in developing policies that would provide the greatest improvement in ecological condition per dollar invested.

At the watershed scale, a key component of risk assessment is targeting major sources of pollution. This trend has emerged from a consensus that many past watershed efforts have taken a fragmented and inefficient approach to controlling NPS pollution (NWQEP, 1988; Water Quality 2000, 1990; U.S. EPA, 1989; Humenick et al., 1987). Efforts to geographically target NPS pollution reduction are based on a growing recognition that small parts of watersheds often contribute a disproportionately large share of NPS pollutants and that these areas need to be targeted to reduce overall watershed pollutant loadings (NWQEP, 1988; CBP, 1990). A good example is the USDA's targeting of highly erodible lands for inclusion in its Conservation Reserve Program. Efforts to improve watershed NPS models should greatly enhance prospects for geographic targeting.

Emphasis on Source Reduction

Increasingly, we are finding that the best way to reduce pollution is not to produce it. As previously discussed, source reduction has proven effective in reducing NPS inputs of lead and organochlorine pesticides. Bans on detergents containing phosphorus have proven to be a cost effective method for reducing phosphorus levels by 50 percent in wastewater effluent. In farming practices, reduction of fertilizer inputs has been identified as a cost-effective approach for reducing nutrients in runoff (Magleby et al., 1990). Approaches to prevent groundwater pollution, such as wellhead protection programs, are vastly less expensive than remediation efforts.

Shift from Volunteer Approaches to Regulatory and Monetary Incentives

Low voluntary participation rates have been a problem in many watershed NPS pollution reduction efforts; voluntary efforts alone are now thought to be insufficient to control major NPS pollution (U.S. EPA, 1989; NWQEP, 1988; CBP, 1990; GAO, 1990). Controls on local land use, animal waste management, and other measures have been suggested as regulatory complements to conventional NPS abatement methods (GAO, 1990; CBP, 1990). Between regulatory control and purely voluntary efforts are quasi-regulatory efforts and cost-share approaches, such as the conservation compliance measures (the "sodbuster" and "swampbuster" provisions) in the 1990 Farm Bill.

There also has been a gradual move from the "command-control" strategy of pollution reduction, which dictates specific pollution reduction methods, toward the use of market incentives. Some market incentive tools, such as the trading of pollution credits, may not be feasible for control of NPS pollution. More realistic may be "pollution taxes" on fertilizers (now in use in Iowa) and pesticides, which would tend to increase the efficiency of application and thereby decrease the likelihood of these chemicals moving offside as pollutants. There has also been considerable interest in eliminating disincentives for reducing agricultural pollution caused by the current crop price support system used in the United States, which tends to maximize crop production at the expense of environmental concerns (NRC, 1989).

USING WETLANDS TO CONTROL NONPOINT SOURCE POLLUTION

The goal of the papers in this volume is to evaluate the prospects for using constructed or natural wetlands to ameliorate the effects of NPS pollution in rural areas. There is relatively little published information on the efficacy of using wetlands to control rural NPS pollution but considerable experience in using wetlands as tertiary treatment for municipal wastewater (Kadlec and Alford, 1989; Knight, 1990; Richardson, 1985). Therefore, a useful starting place in this exercise is to address the question: How is cropland runoff different from the effluent of a secondary wastewater treatment plant, and what are the implications of these differences in considering the potential of wetlands to remove pollutants from cropland runoff? For this comparison, it is useful to employ data from streams in agricultural areas of the "Corn Belt and Dairy Region," which were compiled by Omernik (1976). This region has a high potential for wetlands control of NPS impacts, having both extensive agricultural areas and numerous wetlands and sites suitable for wetland restoration and creation. Data on secondary treatment plant effluent is based on a nationwide compilation of nutrient levels in wastewater treatment plant effluent (Gakstatter et al., 1978).

As a first consideration, note that the chemical composition of secondary effluent from wastewater treatment plants is relatively constant, with standard deviations around 10 percent of the mean (Table 6). Although data are not available for making a similar statistical calculation of uncertainty for cropland runoff, it appears that annual pollutant loadings from cropland may vary by one order of magnitude (reviewed in Novotny and Chesters, 1981). Thus, one could design a wetlands system for tertiary treatment of municipal wastewater from published effluent data, but site-specific loading data would be needed to design an efficient wetland treatment system to treat cropland runoff. Second, whereas total nitrogen concentrations in secondary effluent and cropland runoff are similar, total phosphorus concentrations in cropland runoff are only 1/20 of the concentrations in secondary effluent. Therefore, wastewater effluent is strongly nitrogen limited, with a mean N:P ratio of 2.4:1, whereas cropland runoff is generally phosphorus limited, with a mean N:P ratio of 31:1 (Table 6). Third, both N and P occur primarily in the soluble form in wastewater treatment plant effluent; this is also true for nitrogen in cropland runoff (the predominance of soluble N increases with increasing fertilizer nitrogen use by agriculture; see Omernik, 1977) but not for phosphorus, which occurs primarily in association with particulate matter. Fourth, concentrations of suspended solids are usually <30 mg/L in secondary treatment plant effluent, compared with values of 100-1000 mg/L for cropland. Perhaps the most critical difference, however, is that pollutant loading from croplands is largely event-driven, with extreme variations in both flow and pollutant concentrations, whereas the flow and composition of effluent from a wastewater treatment plant is relatively stable. Concentrations of suspended solids can vary by 2-3 orders of magnitude in cropland runoff within a year, with peak flows carrying the great majority of sediment loading (Figure 8).

Because of these differences, the relative importance of pollution removal processes would be different in wetlands receiving cropland runoff than in wetlands receiving tertiary wastewater. First, sedimentation of particles would be a major process for removing suspended solids and phosphorus in wetlands receiving cropland runoff, since phosphorus is present largely in the particulate form. The high N:P ratios in cropland runoff suggest that plant uptake would be relatively more important as a long-term phosphorus retention process in agricultural wetlands than in wastewater systems, where phosphorus is grossly oversupplied relative to plant nutrient requirements. In contrast, adsorption of phosphorus to soils is a major mechanism of phosphorus removal in wetlands receiving wastewater effluent (Richardson, 1985); sedimentation is unimportant because phosphorus occurs as a soluble species. Denitrification would be an important mechanism in wetlands receiving cropland runoff, perhaps even more efficient than in wastewater wetlands (Kadlec and Alvord, 1989; Knight, 1990), since inorganic nitrogen is oversupplied relative to phosphorus in cropland runoff. A compilation of nutrient balances (Nixon and Lee, 1986) indjcates that typically 20-80 percent of the phosphorus and 10-90 percent of the nitrogen are remained in natural wetlands.

These observations suggest that the design limitation for maximum pollutant removal would be based upon the need to retain sediments during peak flows. In this regard, the design of constructed wetlands for cropland runoff would have more in common with wetlands designed for treatment of urban runoff (Barten, 1986; Meiorin, 1989; Weidenbacher and Willenbring, 1984; also see Loucks, 1990) than with those designed for treatment of secondary wastewater.

In exploring the feasibility of using wetlands for efficient removal of sediment-bound pollutants, the first consideration is the size of wetland needed to control sediment movement. Very limited information from urban systems suggests that the ratio of wetlands to drainage area needed to achieve a reasonable reduction of suspended solids is on the order of 1:20 (Barten, 1986; Meiorin, 1989; Weidenbacher and Willenbring, 1984). A second consideration is the design lifetime. With sedimentation rates in wetlands on the order of 1 cm/yr (McIntyre and Naney, 1991), it appears that the effective lifetime of a wetlands system designed for efficient sediment removal may be on the order of a few decades. Several papers in this volume discuss the additional research needed to develop design criteria for rural wetlands treatment systems.

Pesticides in cropland runoff are also a concern for cropland wetland systems. Of particular interest is atrazine, which is the most heavily used herbicide in the United States. Goolsby and Thurman (1990) showed that for medium-sized watersheds in the Midwest (800-2,000 km^2), the median post-application concentration of atrazine in streams was 10 μg/L, with a 75 percent quartile value of 16 μg/L. This has two implications.

First, atrazine is phytotoxic and is probably toxic to algae at concentrations of 1-10 µg/L (deNoyelles et al., 1982; Kosinski and Merkle, 1984; Johnson, 1986), although it has a fairly short half life (ca. weeks; Huckins et al., 1986). It is, therefore, reasonable to suspect that atrazine would inhibit growth of algae in some wetlands, at least during part of the year. Second, if a constructed wetland were to promote flow to the groundwater system, there would be a possibility of causing groundwater contamination.

In the preceding few paragraphs, wetlands are regarded as engineered systems designed to remove pollutants. There are other benefits to constructing wetlands in rural areas, such as habitat for wildlife. As engineers and wetlands scientists move from using wetlands for tertiary wastewater treatment into the area of rural NPS control, the ancillary benefits of wetland treatment systems may equal or exceed the benefits of the pollutant removal function. Thus, the optimal design of a rural constructed wetland may not necessarily be the one that provides the maximum pollutant removal or even the most cost-effective pollutant removal but the one that balances the pollutant removal function of wetlands with other ecological functions.

REFERENCES

ASIWPCA, 1985. America's Clean Water: The States' Evaluation of Progress, 1972-1982. Association of State and Interstate Water Pollution Control Administrators, Washington, DC.

Baker, D. B., K. A. Krieger, R. P. Richards and J. W. Kramer, 1985a. Gross erosion rates, sediment yields, and nutrient yields for Lake Erie tributaries: Implications for targeting. In: Perspectives on Nonpoint Source Pollution, Proceedings of a National Conference, Kansas City, MO. EPA/440/5-85/001.

Baker, L. A., P. L. Brezonik, and C. Kratzer, 1985b. Nutrient loading models for Florida lakes. pp. 253-258. In: J. F. Taggart and L. M. Moore (eds.), Lake and Reservoir Management, Volume 1. North American Lake Management Society, Washington, DC.

Baker, L. A., J. M. Eilers, R. B. Cook, P. R. Kaufmann, and A. T. Herlihy, 1991. Interregional comparisons of surface water chemistry and biogeochemical processes. pp. 567-613. In: D. F. Charles (ed.), Acidic Deposition and Aquatic Ecosystems: Regional Case Studies. SpringerVerlag, New York, NY.

Barten, J., 1986. Nutrient removal from urban stormwater by wetland filtration: The Clear Lake restoration project. Lake and Reservoir Management, 2: 297-305.

Capel, P. D., 1991. Atmospheric deposition of herbicides on the Mid-Continental United States (abstract). Eos, 71: 1329.

CBP, 1990. Report and Recommendations of the Nonpoint Source Evaluation Panel. Chesapeake Bay Program CBP/TRS 56/91. USEPA, Washington, DC. 28 pp.

CEQ, 1990. Environmental Quality,: 20th Annual Report. Council on Environmental Quality, Executive Office of the President, Washington, DC.

Chiaudani, G. and M. Vighi, 1974. The N:P ratio and tests with Selanastrum to predict eutrophication in lakes. Water Research, 8: 1063-1069.

Deason, J. P., 1989. Impacts of irrigation drainwater on wetlands. pp. 127-138. In: Wetlands: Concerns and Successes. American Water Resources Association, Bethesda, MD.

deNoyelles, R., W. D. Kettle, and D. E. Sinn, 1982. The responses of plankton communities in experimental ponds to atrazine, the most heavily used pesticide in the United States. Ecology, 63: 1285-1293.

Farber, K. D. and G. L. Rutledge, 1988. Pollution abatement and control expenses, 1983-86. Survey of Current Business, May 1988.

Fawcett, R. S., 1987. Overview of pest management for conservation tillage systems. In: T. J. Logan, J. M. Davidson, J. L. Baker, and M. R. Overcash (eds.), Effects of Conservation Tillage on Groundwater Quality: Nitrates and Pesticides. Lewis Publishers, Inc. Chelsea, MI.

Fisher, D., J. Ceraso, T. Mathew, and M. Oppenheimer, 1988. Polluted Coastal Waters: The Role of Acid Rain, Environmental Defense Fund, New York, NY.

GAO, 1990. Greater EPA Leadership Needed to Reduce Nonpoint Source Pollution. GAO/RCED-91-10. GAO, Washington, DC.

Gakstatter, J. H., M. O. Allum, S. E. Dominiguez, and M. R. Crouse, 1978. A survey of phosphorus and nitrogen levels in treated municipal wastewater. Journal of the Water Pollution Control Federation, 50: 718-722.

Goolsby, D. A. and E. M. Thurman, 1990. Heibicides in rivers and streams of the Upper Midwestern United States. In: Proceedings of the Forty-sixth Annual Meeting of the Upper Mississippi River Conservation Committee, Bettendorf, IA.

Grover, R, 1991. Nature, transport, and fate of airborne residues. pp. 90-117. In: R. Grover and A. J. Cessna (eds.), Environmental Chemistry of Herbicjdes, Vol. 2. CRC Press, Boca Raton, FL.

Herlihy, A. T., P. R. Kaufmann, and M. E. Mitch, 1990. Regional estimates of acid mine drainage impact on streams in the mid-Atlantic and southeastern United States. Water, Air, and Soil Pollution, 50: 91-107.

Hindall, S. M., 1975. Measurement and Prediction of Sediment Yields in Wisconsin Streams. Water Resources Investigation Report, U.S. Geological Survey, Reston, VA. pp. 54-75.

Hren, J., C. J. O. Childress, J. M. Norris, T. H. Chaney, and D. N. Myers, 1990. Regional water quality: Evaluation of data for assessing conditions and trends. Environmental Science and Technology, 24: 1122-1127.

Huckins, J. N., J. D. Petty, and D. C. England, 1986. Distribution and impact of trifluralin, atrazine, and fonofos residues in microcosms simulating a northern prairie wetland. Chemosphere, 15: 563-588.

Hughes, R. M. and D. P. Larsen, 1988. Ecoregions: an approach to surface water protection. Journal of the Water Pollution Control Federation, 60: 486-493.

Humenik, F. J., M. D. Smolen, and S. A. Dressing, 1987. Pollution from nonpoint sources. Environmental Science and Technology, 21: 737-742.

Johnson, B. T., 1986. Potential impact of selected agricultural chemical contaminants on a northern prairie wetland: a microcosm evaluation. Environmental Toxicology and Chemistry, 5: 473-485.

Kadlec, R. H. and H. Alvord, 1989. Mechanisms of water quality improvement in wetland treatment systems. In: D. W. Fisk (ed.), Wetlands: Concerns and Successes. American Water Resources Association, Bethesda, MD.

Knight, R. L., 1990. Wetland systems. In Natural Systems for Wastewater Treatment, Manual of Practices FD-16. Water Pollution Control Federation.

Kosinksi, R. J. and M. G. Merkle, 1984. The effect of four terrestrial herbicides on the productivity of artificial stream algal communities. Journal of Environmental Quality, 13: 75-82.

Larsen, D. P., D. R. Dudley, and R. M. Hughes, 1988. A regional approach to assess attainable water quality: An Ohio case study. Journal of Soil and Water Conservation, 43: 171-176.

Lee, L. K., 1990. The dynamics of declining soil erosion rates. Journal of Soil and Water Conservation, 45: 622-624.

Lettenmaier, D. P., E. R. Hooper, C. Wagoner, and K. B. Faris, 1991. Trends in stream water quality in the continental United States, 1978-1987. Water Resources Research, 27: 327-340.

Logan, T. J., 1987. An assessment of Great Lakes tillage practices and their potential impact on water quality. In: T. J. Logan, J. M. Davidson, J. L. Baker, and M. R. Overcash (eds.), Effects of Conservation Tillage on Groundwater Quality: Nitrates and Pesticides. Lewis Publishers, Inc., Chelsea, MI.

Loucks, O. L., 1990. Restoration of the pulse control function of wetlands and its relationship to water quality objectives pp. 467-477. In: J. A. Kusler and M. E. Kentula (eds.), Wetland Creation and Restoration: The Status of the Science. Island Press, Washington, DC.

Magleby, R. S., S. Piper, and C. E. Young, 1990. Economic insights on nonpoint pollution control and the Rural Clean Water Program. pp. 63-69. In: Making Nonpoint Pollution Control Work: Proceedings of a National Conference, St. Louis, MO. National Association of Conservation Districts, League City, TX.

McIntyre, S. C. and J. W. Naney, 1991. Sediment deposition in a forested inland wetland with a steep-farmed watershed. Journal of Soil and Water Conservation, 46: 64-66.

Meade, R. H. and R. S. Parker, 1985. Sediments in Rivers of the United States. National Water Summary, 1984. Water Supply Paper 2275. U.S. Geological Survey, Reston, VA.

Meiorin, E. C., 1989. Urban runoff treatment in a fresh/brackish water marsh in Fremont, California pp. 677-685. In: D. A. Hammer (ed.), Constructed wetlands for Wastewater Treatment: Municipal, Industrial, and Agricultural. Lewis Publishers, Inc., Chelsea, MI.

Miller, W. E., T. E. Maloney, and J. C. Greene, 1974. Algal productivity in 49 lake waters as determined by algal assays. Water Research, 8: 667-679.

Moore, J. N. and S. N. Looma, 1990. Hazardous wastes from large-scale metal extraction: A case study. Environmental Science and Technology, 24: 1278-1285.

Nielsen, E. G. and L. K. Lee, 1987. The Magnitude and Costs of Groundwater Contamination from Agricultural Chemicals: A National Perspective, AGES870318. National Resources Economics Division, Economic Research Service, USDA, Washington, DC.

Nixon, S. W. and V. Lee, 1986. Wetlands and Water Quality: A Regional Review of Recent Research in the United States on the Role of Freshwater and Saltwater Wetlands as Sources, Sinks, and Transformers of Nitrogen, Phosphorus, and Various Heavy Metals. Technical Report Y-86-2, U.S. Army Corps of Engineers, Washington, DC. 229 pp.

Novotny, V. and G. Chesters, 1981. Handbook of Nonpoint Pollution: Sources and Management. Van Nostrand Reinhold Company, New York, NY.

NRC, 1989. Alternative Agriculture. Committee on the Role of Alternative Farming Methods in Modern Production Agriculture, Board on Agriculture, National Research Council, National Academic Press, Washington, DC. 448 pp.

NWQEP, 1988. 1987 Annual Report: Status of Agricultural Nonpoint Source Projects. National Water Quality Evaluation Project, Office of Water Regulations and Standards Division, USEPA, IWH-585, Washington, DC.

Omernik, J. M., 1976. The Influence of Land Use on Stream Nutrient Levels. EPA/600/3-76/014. USEPA Environmental Research Laboratory, Corvallis, OR.

Omernik, J. M., 1977. Nonpoint Source-Stream Nutrjent Level Relationships: A Nationwide Survey. EPA/600/3-77/105. USEPA Environmental Research Laboratory, Corvallis, OR.

Otterby, M. A., and C. A. Onstad, 1981. Average Sediment Yields in Minnesota. ARR-NC-8. USDA Agricultural Research Service. 9 pp.

Rappaport, R. A., N. R. Urban, P. D. Capel, J. B. Baker, B. Looney, and S. J. Eisenreich, 1984. "New" DDT inputs to North America: atmospheric deposition. Chemosphere, 14: 1167-1173.

Richardson, C. J., 1985. Mechanisms controlling phosphorus retention capacity in freshwater wetlands. Science, 228: 1424-1427.

SAB, 1990. Reducing Risk: Setting Priorities and Strategies for Environmental Protection. SABEC-90-021. Scientific Advisory Board, USEPA, Washington, DC.

Schmitt, C. J. and W. G. Brumbaugh, 1990. National contaminant biomonitoring program: concentrations of arsenic, cadmium, lead, mercury, selenium, and zinc in U.S. freshwater fishes, 1976-1984. Archives of Environmental Toxicology, 19: 731-747.

Schmitt, C. J., J. L. Zajicek, and P. H. Peterman, 1990. National contaminant biomonitoring program: residues of organochlorine chemicals in U.S. freshwater fish, 1976-1984. Archives of Environmental Toxicology, 19: 748-781.

Smith, R. A., R. B. Alexander, and M. G. Wolman, 1987. Water quality trends in the nation's rivers. Science, 235: 1608-1615.

Stoddard, J., 1991. Aquatic effects of nitrogen oxides. pp. 10-120 - 10-239. In: Air Quality Criteria for Oxides of Nitrogen: External Review Draft, August, 1991. EPA/600/8/91/0496-A. USEPA.

Treffry J. H., S. Metz, R. P. Trocine, and T. A. Nelson, 1985. A decline in lead transport by the Mississippi River. Science, 230: 439-441.

Tyler, M. 1988. Contribution of Atmospheric Nitrate Deposition to Nitrate Loading in the Chesapeake Bay. VERSAR, Inc., Columbia, MD.

Trimble, S. W., 1981. Changes in sediment storage in the Coon Creek basin, Driftless Area, Wisconsin, 1853-1975. Science, 214: 181-183.

U.S. Department of Agriculture, 1987. Agricultural Resources: Cropland, after and Conservation Situation and Outlook Report. Economic Research Service, USDA, Washington, DC. 39pp.

U.S. Department of Agriculture, 1990. Agricultural Resources: Cropland, Water and Conservation Situation and Outlook Report. Economic Research Service, USDA, Washington., DC. 55 pp.

U.S. Department of the Interior, 1989. Quality of Water: Colorado River Basin, Progress Report No. 14, USDI.

U.S. Environmental Protection Agency, 1983. Results of the Nationwide Urban Runoff Program, Volume 1 - Final Report. WH-554. Water Planning Division, USEPA, Washington, DC.

U.S. Environmental Protection Agency, 1986. National Water Quality Inventory: Report to Congress. USEPA, Washington, DC.

U.S. Environmental Protection Agency, 1989. Nonpoint Sources: Agenda for the Future. WH556. USEPA, Office of Water, Washington, DC. 31pp.

U.S. Environmental Protection Agency, 1990a. National Water Quality Inventory: 1988 Report to Congress. EPA/440/4-90/003. USEPA, Washington, DC.

U.S. Environmental Protection Agency, 1990b. Managing Nonpoint Source Pollution: Final Report to Congress on Section 319 of the Clean Water Act (1989), EPA/506/9-90. USEPA, Washington, DC.

U.S. Environmental Protection Agency 1990c. National Pesticide Survey: Project Summary. USEPA, Washington, DC.

U.S. Environmental Protection Agency, 1991. EMAP Monitor, January 1991. Office of Research and Development, USEPA, Washington, DC.

Van der Leaden, F., F. O. Troise, and D. K. Todd, 1990. The Water Encyclopedia. 2nd edition. Lewis Publishers, Inc., Chelsea, MI.

Water Quality 2000, 1990. Water Quality 2000: Phase II Report, Problem Identification. Water Quality), 2000, Alexandria, VA.

Weidenbacher, W. D. and P. R. Willenbring, 1984. Limiting nutrient flux into an urban lake by natural treatment and diversion. Lake and Reservoir Management, 3: 525-526.

Williams, W. M., P. W. Holdren, D. W. Parsons, and M. N. Lorber, 1988. Pesticides in Ground Qater Data Base: 1988 Interim Report. Office of Pesticide Programs, USEPA, Washington, DC.

Table 1. Attainment of Designated Use for U.S. Surface Waters
Source: U.S. EPA, 1990a

	Rivers and streams*	Lakes&	Estuaries and coastal waters$
Total national resource	1,800,000	39,400,000	35,198
Resource within assessed states	1,150,482	22,347,961	not given
Assessed resource	519,413	16,314,012	26,676
Assessed waters meeting use designation, %:			
Fully supporting	69.6	73.7	71.6
Threatened:#	6.9	17.8	1.3
Partially supporting	20.1	16.6	22.8
Not supporting	10.3	9.8	5.6

* Resource estimates are in miles. The total resource value is from ASIWPCA (1985). The assessed resource is for 48 reporting states only.

& Resource estimate in acres. No primary reference given for the total lake resource. The assessed resource is for 40 reporting states.

$ Resource estimates in square miles; 23 out of 30 estuarine states reporting.

The "threatened" category is a subset of "fully supporting" in the 305(b) report. Thus, percentages sum to more than 100%.

Table 2. **Water Quality Impairment by Nonpoint Source Pollutants Source: U.S. EPA, 1990b**

	Rivers[*]	Lakes[&]	Estuaries[$]
Nonpoint source impacts reported	206,179	5.3×10^6	5,800
Degree of impact, %[#]			
Nonsupport	52	42	54
Partial support	28	22	36
Threatened	20	36	10
Causes, %			
Agriculture	41	23	7
Urban	4	6	11
Land disposal	3	4	8
Construction	2	2	--
Hydromodification	6	6	--
Mining	8	7	16
Natural	8	10	--
Others	3	16	43
In-place	--	--	16
Unknown	23	21	4

[*] Resource estimate in miles. 40 states reporting impacts; 20 reported identified use impairment; 33 identified causes of NPS pollution.

[&] Resource estimates in acres. The number of states reporting impacts is not given; 18 states identified the type of-use impairment; 25 states identified causes of NPS pollution.

[$] Resource estimate in square miles. Thirteen states reporting impacts.

[#] EPA (1990b) includes "threatened" in the "impaired" category, in contrast to the 305(b) report (EPA 1990a).

Table 3. **Trends of Selected Water Quality Variables in Major Rivers of the United States**
Sources: Smith et al., 1987 (for 1974-1981); Lettenmaier et al., 1991 (for 1978-1987)

		1974-1981			1978-1987		
		Number of stations			Number of stations		
Constituent	**Median***	**Increasing**	**Decreasing**	**No Change**	**Increasing**	**Decreasing**	**No Change**
Chloride	15	101	34	154	65	32	295
Suspended solids	67	43	39	194	12	19	121
Total P	0.13	43	50	288	12	69	308
Nitrate**	0.41	116	27	240	82	24	284
Lead	4	2	23	219	2	23	325
Cadmium	< 2	16	2	231	1	16	298
Arsenic	< 1	21	4	220	1	25	286

* Median concentrations are in mg/L for major common constituents and µg/L for trace metals (lead, cadmium, and arsenic).

** 1978-1987 data are for total nitrogen.

Table 4. **Nitrogen Budget for the Chesapeake Bay***
Source: Stoddard (1991), modified from Tyler (1988) and Fisher et al. (1988)

	Inputs to the watershed	Flux to the Bay
Fertilizer	11.3	0.4
Animal waste	13.9	0.4
Nitrogen deposition to land	11.4	1.1
Total for watershed	36.6	1.9
Point sources	-	2.4
Nitrogen deposition to bay surface	-	0.6
Total nitrogen input:		4.9

* Inputs and fluxes are in 10^9 moles/year.

Table 5. Pesticides and Nitrate in Community Well Systems and Rural Domestic Wells Source: U.S. EPA, 1990c*

		% Detected			% above HAL/MCL***	
Constituent**	**Use**	**CWS**	**RDW**	**HAL/MCL**	**CWS**	**RDW**
Nitrate	**Fertilizer**	**52.1**	**57.0**	**10**	**1.2**	**2.4**
Pesticides	**Various**	**10.4**	**4.2**	**Various**	**0**	**0.6**
Atrazine	Herbicide	1.7	0.7	3	--	X
DBCP	Nematode	0.4	0.4	0.2	--	X
Alachlor	herbicide	0	0.03	2	--	X
EDB	Insecticide	0	0.2	0.05	--	X
Lindane	Insecticide	0	0.1	0.2	--	X
DCPA	Herbicide	6.4	2.5	4,000	--	--
Prometon	?	0.5	2.6	100	--	--
Simazine	?	1.1	0.2	1.0	--	--
Hexachlorobenzene	Fungicide	0.5	0	1	--	--
Dinoseb	Herb./Fung.	0.03	0	7	--	--
Bentazon	Herbicide	0	0.1	20	--	--

* This was a statistically based survey of community water supply (CWS) wells and rural domestic wells (RDW). During 1988-1990, 1,300 wells were sampled, representing an estimated population of 94,600 community water supply wells and 10.5 million rural domestic wells.

** Pesticide acroniyms are as follows: EDB = ethylene dibromide; DBCP = 1,2-dibromo-3-chloropropane; DCPA = dimethyl tetrachloroterephthalate.

*** For individual pesticides, percentages above HAL/MCL are not given. An "X" indicates that some wells exceeded the HAL/MCL and a "--" indicates that none did. HAL = health advisory limits; MCL = maximum contaminant levels.

Table 6. Characteristics of Secondary Effluent from Municipal Wastewater Treatment Plant, and Cropland Runoff

	Secondary effluent[*]	Cornbelt cropland[&]	
	Concentration (mg/L)	Concentration (mg/L)	Loading hg/ha/yr
Suspended solids[&]	10-30	50-1000	
Total P	6.8 ± 0.4	0.14	0.31
Soluble P	5.3 ± 0.4	0.06	0.13
Total N	15.8 ± 1.2	4.4	9.5
Sol. inorg. N	8.4 ± 0.45	3.4	7.3
N:P	2.4	31.4	

[*] From Gakstatter et al., 1978. Means and standard deviations represent effluent from 244 activated sludge treatment plants located throughout the country.

[&] Source: Omernik, 1976. Total P and N loadings are means for "mostly agricultural" land (> 90% cropland) within the "Corn Belt and Dairy" region, based upon 80 sampled streams. Soluble P and N are calculated as the product of the ratio of soluble/total nutrients, calculated for "mostly agricultural" lands throughout the country (n = 91), and the total nutrient concentrations in the Corn Belt and Dairy region.

[$] The range for suspended solids is from Table 1-3 of Novotny and Chesters (1981). The estimate for cropland is a range for stream yields in the agricultural areas of Wisconsin and Minnesota (Otterby and Onstad, 1981; Hindall, 1975).

LIST OF FIGURES

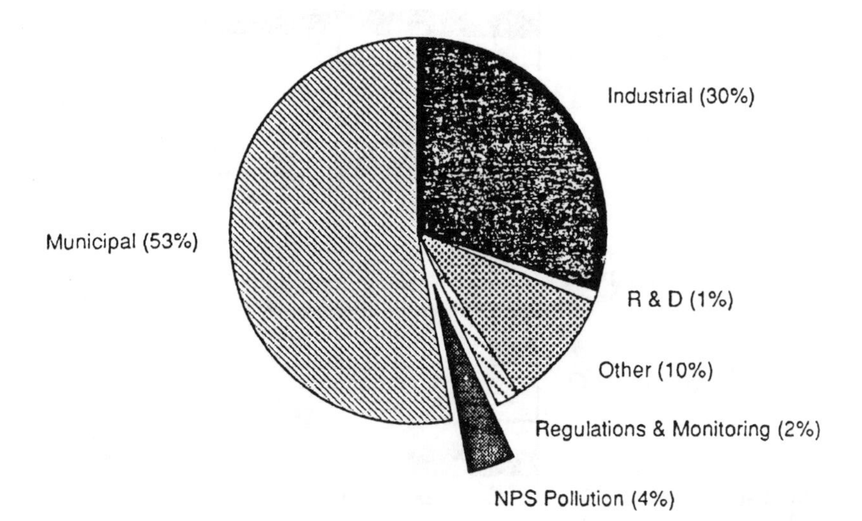

Figure 1

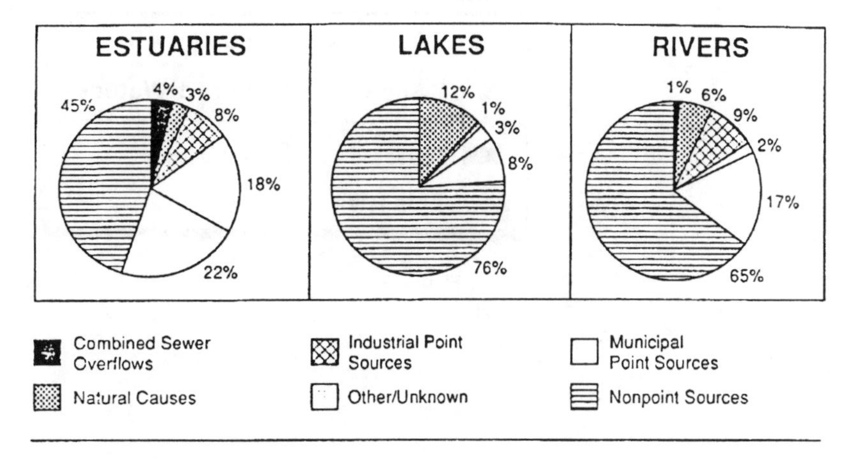

Figure 2

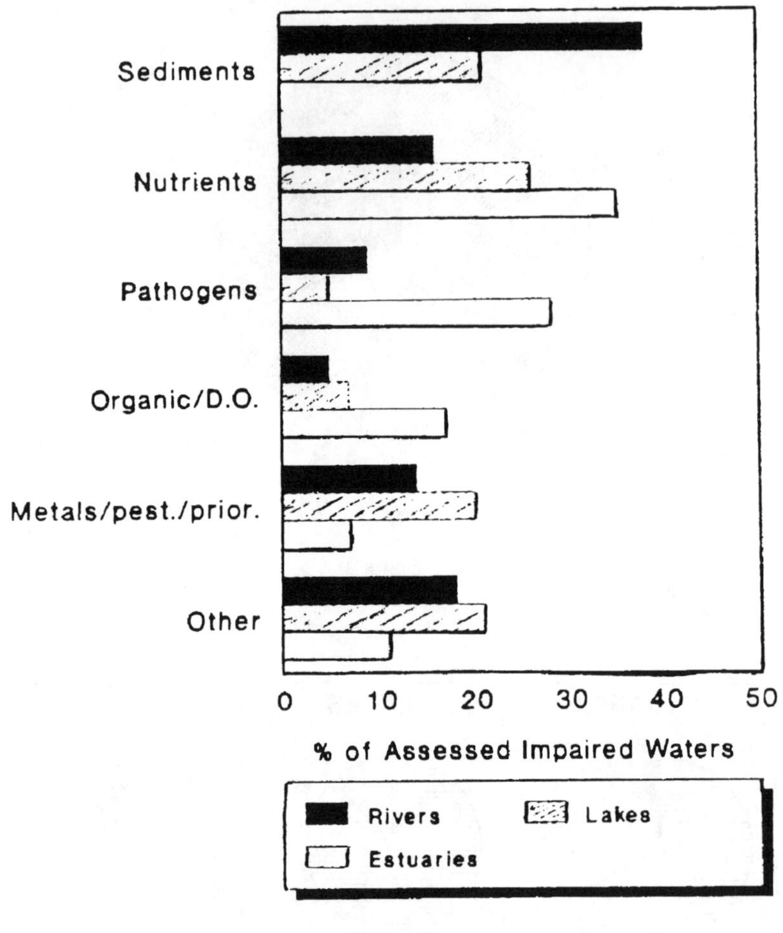
Sediments
Nutrients
Pathogens
Organic/D.O.
Metals/pest./prior.
Other
0
10
20
30
40
50
% of Assessed Impaired Waters
Rivers
Lakes
Estuaries

Figure 3

A. 1853-1938

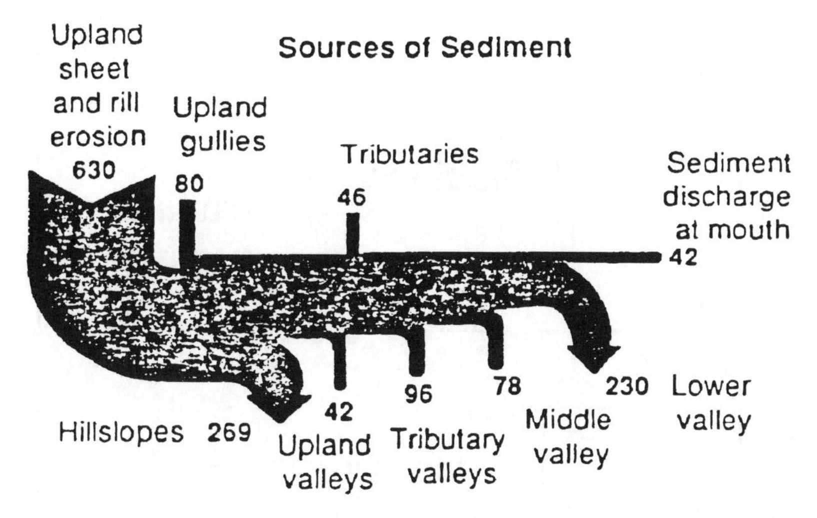

B. 1938-1975

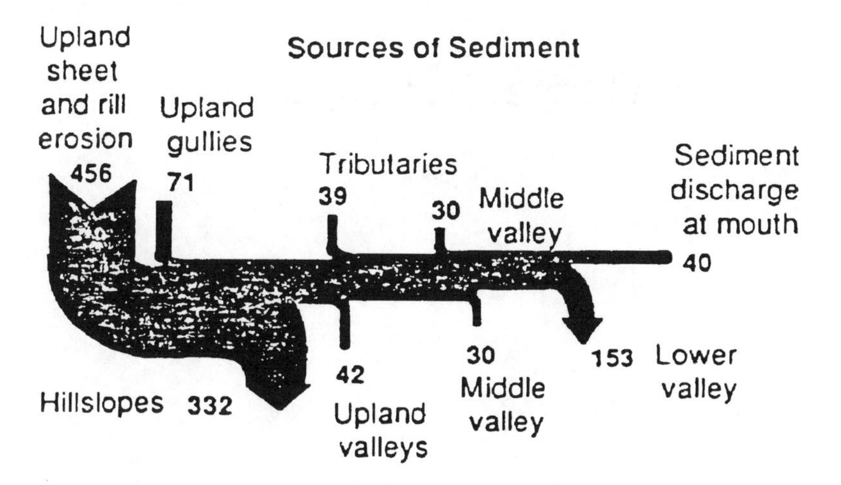

Figure 4

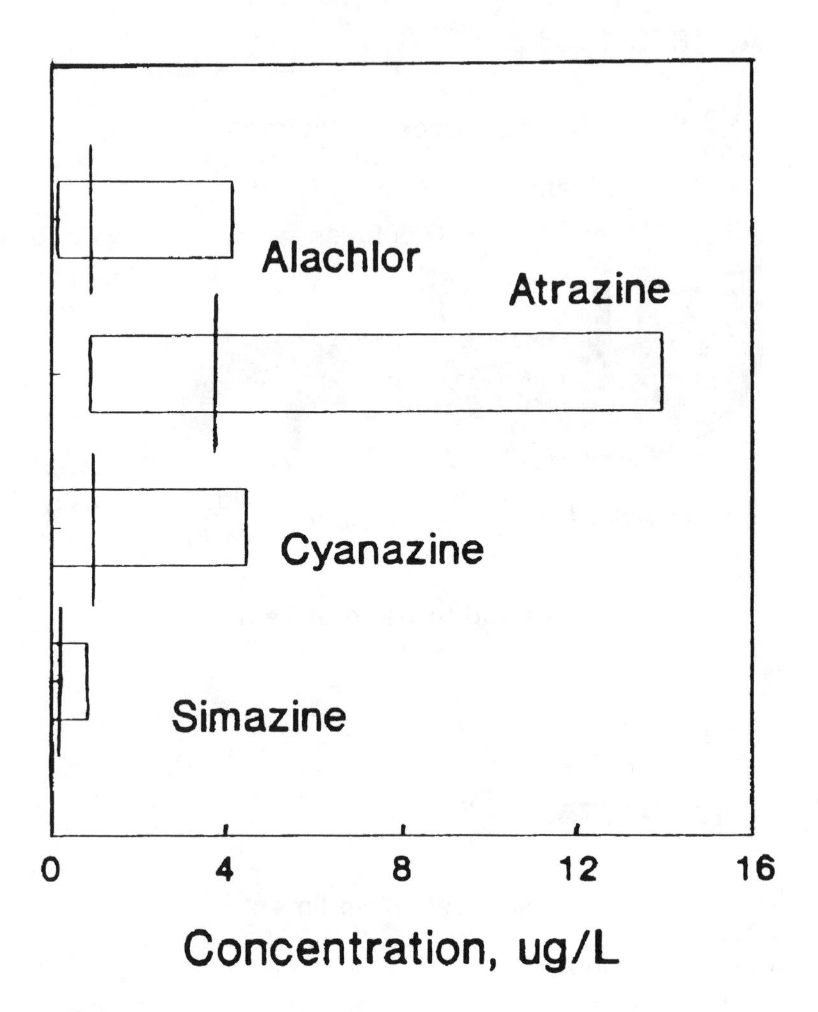
Alachlor
Atrazine
Cyanazine
Simazine
0
4
8
12
16
Concentration, ug/L

Figure 5

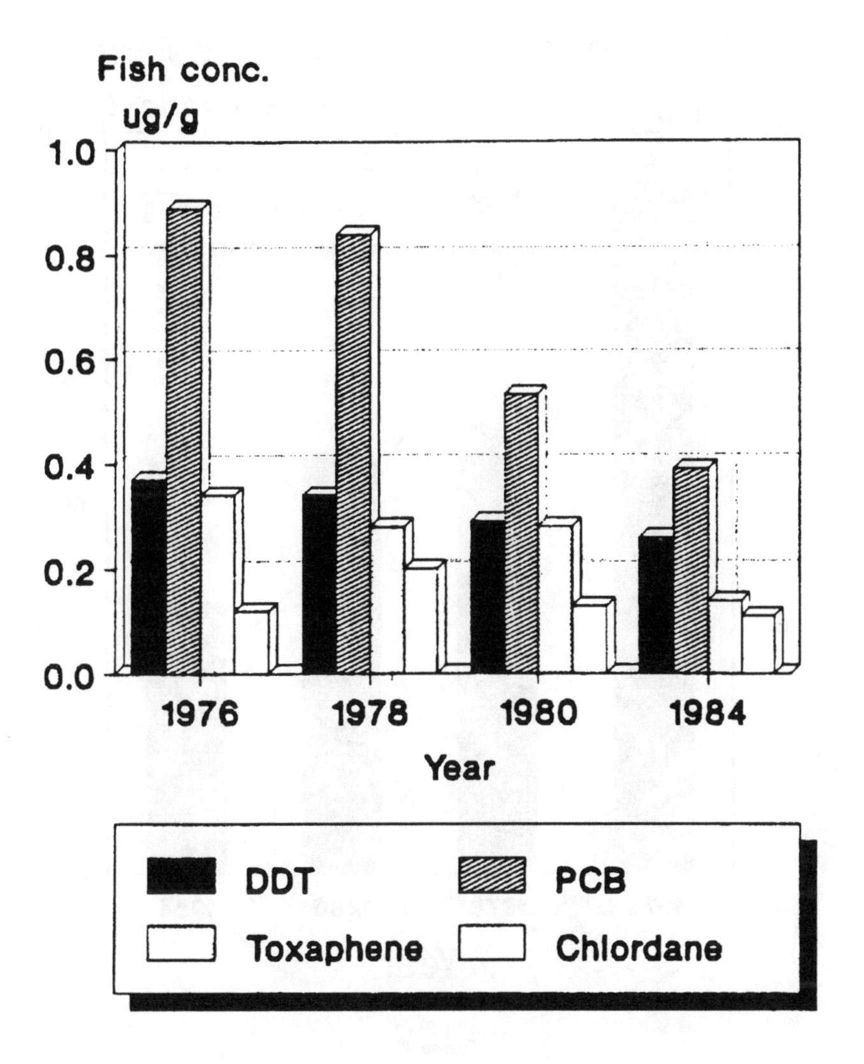
Fish conc.
ug/g
1.0
0.8
0.6
0.4
0.2
0.0
1976
1978
1980
1984
Year
DDT
PCB
Toxaphene
Chlordane

Figure 6

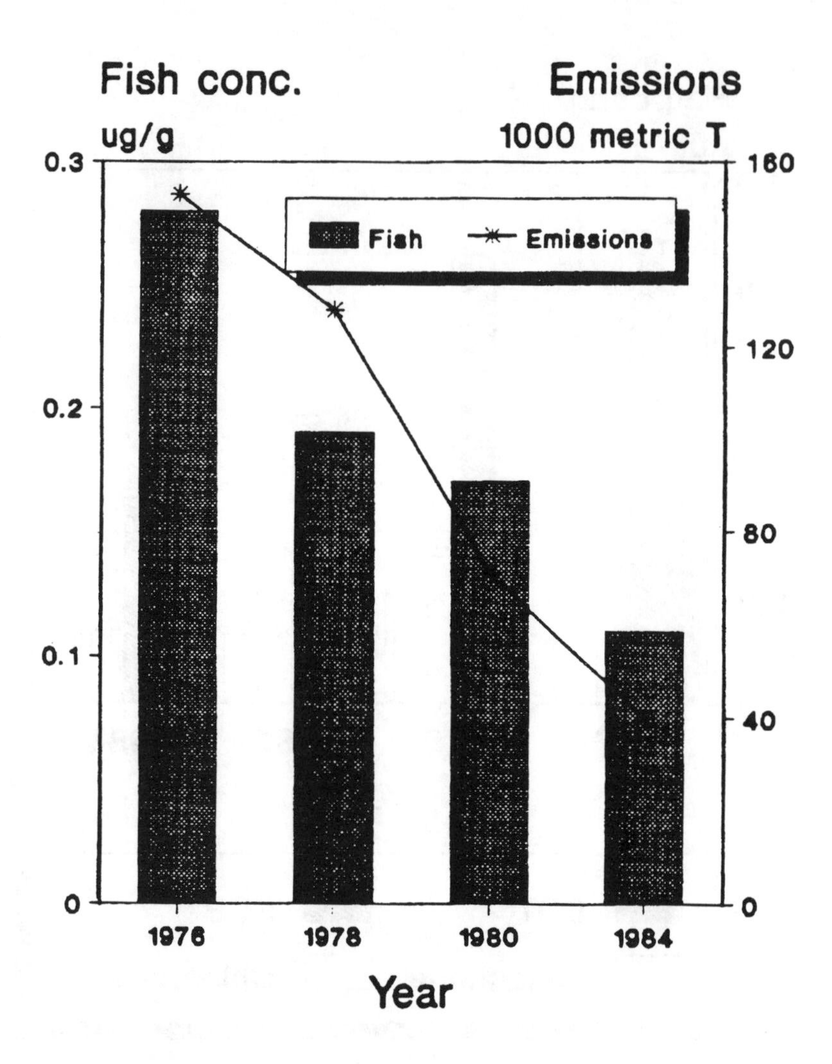
Fish conc.
ug/g
Emissions
1000 metric T
0.3
0.2
0.1
0
160
120
80
40
0
Fish
Emissions
1976
1978
1980
1984
Year

Figure 7

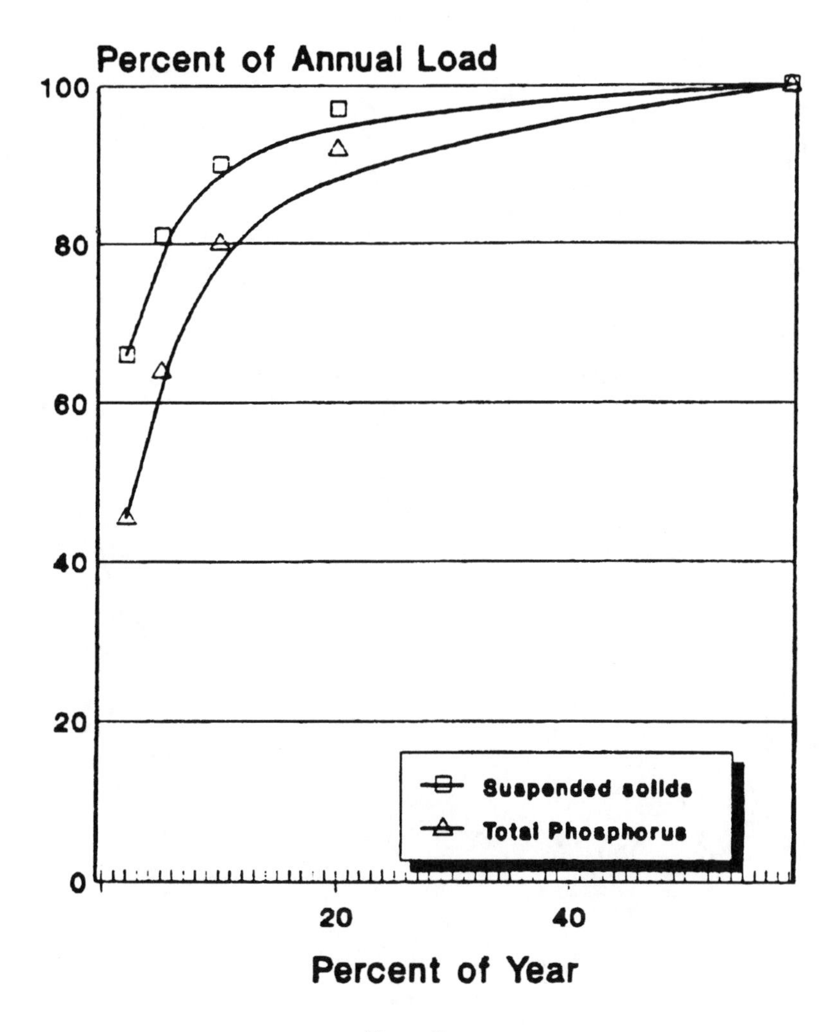
Percent of Annual Load
100
80
60
40
20
0
20
40
Percent of Year
Suspended solids
Total Phosphorus

Figure 8

CHAPTER 3

Landscape Design and the Role of Created, Restored, and Natural Riparian Wetlands in Controlling Nonpoint Source Pollution

William J. Mitsch, School of Natural Resources

ABSTRACT

General design principles, landscape locations, and case studies of natural and constructed riverine wetlands for the control of nonpoint source (NPS) water pollution are presented. General design principles of wetland construction for NPS pollution control emphasize self-design and minimum maintenance systems, with an emphasis on function over form and biological form over rigid designs. These wetlands can be located as instream wetlands or as floodplain riparian wetlands, can be located as several wetlands in upstream reaches or fewer in downstream reaches of a watershed, and can be designed as terraced wetlands in steep terrain-Case studies of a natural riparian wetland in southern Illinois, an instream wetland in a downstream location in a northern Ohio watershed, and several constructed riparian wetlands in northeastern Illinois demonstrate a wide range of sediment and phosphorus retention, with greater efficiencies generally present in the constructed wetlands (63-96 percent retention of phosphorus) than in natural wetlands (4-10 percent retention of phosphorus). By itself, this could be misleading as the natural wetlands have much higher loading rates and actually retain an amount of nutrients comparable to constructed wetlands (1-4 $_{g-}$P $m^{-2}4^{-1}$).

INTRODUCTION

The riparian wetlands of the United States, which once were connected to many of the streams, rivers, and Great Lakes of the Nation, have for all intents and purposes been eliminated from the landscape. With the wetland loss and their conversion to other land uses, our rivers and streams no longer have the ability to cleanse themselves, and bodies of water such as the Great Lakes are no longer buffered from upland regions. The net result has been poorer water quality, particularly in the glaciated Midwest where nonpoint source (NPS) pollution in now pervasive. In midwestern states such as Ohio, Indiana, Illinois, and Iowa, where more than 80 percent of the wetlands have been drained, partially in response to the Swamp Lands Acts of 1849, 1850, and 1860, water quality has been particularly degraded. With a 90 percent loss of wetlands, Ohio is ranked below only California for the highest percent loss of wetlands from the 1780s to the 1980s (Dahl, 1990). NPS water pollution problems are common in Ohio, and nutrients continue to flow into Lake Erie to the north and the Ohio River to the south.

Much national attention is now focused on the design and construction of wetlands. In 1987, a National Wetlands Policy Forum was convened by the Conservation Foundation at the request of the U.S. Environmental Protection Agency (EPA) to investigate the issue of wetland management in the United States (NWPF, 1988; Davis, 1989), The distinguished group of 20 members-including three governors, a State legislator, State and local agency heads, chief executive officers of environmental groups and businesses, farmers, ranchers, and academic experts-published a report that set significant goals for the Nation's remaining wetlands. The report recommended a policy "to achieve no overall net loss of the nation's remaining wetlands base and to create and restore wetlands, where feasible, to increase the quantity and quality of the nation's wetland resource base" (NWPF, 1988).

In his 1990 budget address to Congress, President Bush echoed the "no net loss" concept as a national goal (Davis, 1989), shifting the activities of many agencies such as the Department of Interior, the EPA, the U.S. Army Corps of Engineers, and the Department of Agriculture to provide leadership toward a unified and seemingly simple goal. It was not anticipated that there would be a complete halt of wetland loss in the United States when economic or political reasons dictated otherwise, so implied in the "no net loss" concept is wetland construction and restoration.

This paper will present some general principles of ecological engineering for the proper design of wetlands, some design ideas for the placement of wetlands in the landscape, and case studies of three riparian wetland systems that have been evaluated for their role in controlling NPS pollution.

ECOLOGICAL ENGINEERING OF WETLANDS

Ecological engineering combines basic and applied science for the restoration, design, and construction of ecosystems, including wetlands (see Mitsch, 1988, 1991; Mitsch and Jorgensen, 1989 for details), The goals of ecological engineering and ecotechnology are: (1) the restoration of ecosystems that have been substantially disturbed by human impacts such as environmental pollution, climate change, or land disturbance; (2) the development of new sustainable ecosystems that have both human and ecological value; and (3) the identification and protection of the lifesupport value of existing ecosystems. Combining ecosystem function with human needs is the foundation of ecological engineering (also called "ecotechnology"), presently defined as "the design of human society with its natural environment for the benefit of both" (Mitsch and Jorgensen, 1989). Ecological engineering provides approaches for conserving our natural environment while at the same time adapting to and sometimes solving intractable environmental pollution problems.

CREATING AND RESTORING WETLANDS

The creation and restoration of wetlands need to be accomplished in ecologically sound ways. Wetland creation refers to the construction of wetlands where they did not exist before. These *created wetlands* are also called *constricted wetlands* or *artificial wetlands*, although the last term is not preferred by many wetland scientists. No one has estimated the number of such wetlands in the United States, but it is probably in the thousands. *Wetland restoration* refers to the enhancement of existing wetlands, often with only the hydric soils as an indicator of former wetlands.

Wetlands are constructed or restored in the landscape for a variety of reasons or objectives. The three most popular reasons for wetland construction in the United States have been to treat wastewater, to compensate a wetland loss elsewhere (mitigation), and to provide habitat for wildlife. A more recent application under study in a number of locations around the country is to control NPS runoff, especially from rural agricultural lands. Over the past 15 years, many natural wetlands have been evaluated for that role as well. Wetlands built for the control of NPS pollution (e.g., sediments and nutrients) are often part of a watershed or river floodplain restoration project (Hey et al., 1989; Livingston, 1989; Newton, 1989; Mitsch, 1990; Mitsch and Cronk, in press). The construction of wetlands for NPS pollution control contributes to two national goals-cleaning up our Nation's waterways and adding to our Nation's wetlands reserves.

Preliminary Principles of Wetland Design

Some of the principles of ecological engineering that could be applied to the construction and restoration of wetlands for NPS pollution control are outlined below:

1. Design the system for minimum maintenance. The system of plants, animals, microbes, substrate, and water flows should be developed for self-maintenance and self-design (Mitsch and Jorgensen, 1989; Odum, 1989).
2. Design a system that utilizes natural energies, such as potential energy of streams, as natural subsidies to the system. Pulsing streams during Midwestern spring transport great quantities of nutrients in relatively short periods.
3. Design the system *with* the landscape, not against it. Floods and droughts are to be expected, not feared. Outbreak of plant diseases and invasion of alien species are often symptomatic of other stresses and may indicate faulty design rather than ecosystem failure.
4. Design the system with multiple objectives, but identify at least one major objective and several secondary objectives.
5. Design the system as an ecotone. This means including a buffer strip around the site, but it also means that the wetland site itself is often a buffer system between upland and aquatic systems.
6. Give the system time. Wetlands do not become functional overnight and several years may elapse before nutrient retention or wildlife enhancement is optimal. Strategies that try to short-circuit ecological succession or over-manage are doomed to failure.
7. Design the system for function, not for form. If initial plantings and animal introductions fail but the overall function of the wetland-based on initial objectives-is intact, then the wetland has not failed. Expect the unexpected.

8. Do not over-engineer wetland design with rectangular basins, rigid structures and channels, and regular morphology. Ecological engineering recognizes that natural systems should be mimicked to accommodate biological systems (Brooks, 1989).

LOCATING WETLANDS IN THE LANDSCAPE

Wetlands are effective in retaining some nutrients and sediments from NPS pollution. Wetlands should not be expected, however, to control ali of the influx of sediments and nutrients from a watershed, nor should the creation of one small wetland be expected to result in significant improvements in downstream water quality. If wetlands are to be constructed in the watershed for the control of NPS pollution, a variety of possible designs should be considered.

Instream Wetlands

Wetlands can, of course, be designed as instream systems by adding control structures to the streams themselves or by impounding a distributary of the stream (Figure 1). Blocking an entire stream is a reasonable alternative only in low-order streams, and it is not generally cost-effective. This design is particularly vulnerable during flooding and might be very unpredictable in its ultimate stability. It has the advantage, however, of potentially "treating" a significant portion of the water that passes that point in the stream. Maintenance of the control structure and the distributary might mean significant management commitments to this design.

Riparian Wetlands

The natural design for a riparian wetland primarily fed by a flooding stream (Figure 2) allows for flood events of a river to deposit sediments and chemicals on a seasonal basis in the wetland. Because there are both man-made and natural levees along major sections of streams, it is often possible to create such a wetland with minimal construction. The wetland could be designed to capture flooding water and sediments and slowly release the water back to the river after the flood passes. This is the design of natural riparian wetlands in bottomland hardwood forest areas (e.g., Mitsch et al., 1979). The wetland could also be designed to receive water from flooding and retain it through the use of flap-gates.

A riparian wetland fed by a pump (Figure 3) creates the most predictable hydrologic conditions for the wetland, but at an extensive cost for equipment and maintenance. This design is being used for some of the wetlands at the Des Plaines River Demonstration Project (see below), but for the specific purpose of establishing experimental conditions for research. If it is anticipated that the primary objective for building a constructed wetland is the development of a research program to determine design parameters for future wetland construction in the basin, then a series of wetlands fed by pumps is a good design. If other objectives are most important, then the use of large pumps is usually not appropriate. Small pumps may be necessary to carry wetlands through drought periods.

In a compromise between the designs in Figures 2 and 3, a wetland could be fed by diversion of the stream at such a distance upstream of the wetland that it could effectively be fed by gravity (Figure 4). In such a design, natural energies rather than pumps could be used, and natural levees rather than impoundments would effectively hold the water on the floodplain.

Upstream Versus Downstream Wetlands

The advantages of locating several small wetlands in the upper reaches of a watershed (but not in the streams themselves) rather than fewer larger wetlands in the lower reaches should be considered (Figure 5). Loucks (1989) argues that locating a greater number of low-cost wetlands in the upper reaches of a watershed rather than building fewer high-cost wetlands in the lower reaches offers a better strategy for wetlands to survive extreme events. However, a modeling effort on flood control by Ogawa and Male (1983) suggested the opposite: the usefulness of wetlands in decreasing flooding increases with the distance the wetland is downstream. Figure 6 shows a design where multiple wetlands are constructed in the landscape to intercept small streams and drainage tiles. The main stream itself is not diverted, but the wetlands receive their water, sediments, and nutrients from small tributaries, swales, and overland flow. More significantly, if tile drains can be located and broken upstream of their discharge into tributaries, they can be very effective conduits for supplying adequate water to the wetlands. Because these tile drains are often the sources of the highest concentrations of chemicals such as nitrates from agricultural fields, the wetlands could be very effective in controlling certain types of NPS pollution while creating a needed habitat in an agricultural setting.

Wetlands in Steep Terrain

Wetlands are a phenomenon of naturally flat terrain. However, steeper terrain is often most susceptible to high erosion and hence high contributions of suspended sediments and other chemicals. One approach is to attempt to integrate "terraced" wetlands into the landscape (Figure 7). In this case, wetland basins are constructed as smaller basins that "stair-step" down steep terrain. While there are some examples of these types of wetlands, particularly in the building of acid mine drainage wetlands in the Appalachian mountains, few wetlands of this type have been constructed for the control of NPS pollution.

CASE STUDIES-WETLANDS THAT CONTROL NONPOINT SOURCE POLLUTION

Heron Pond, Illinois – a Natural Riparian Wetland

Early work by Mitsch et al. (1977, 1979) on a riparian forested wetland in southern Illinois demonstrated the ability of these systems to serve as nutrient and sediment traps. Heron Pond is a 30 ha alluvial cypress swamp dominated by bald cypress (Taxodium distichum [L.] Rich.) and water tupelo (Nyssa aquafica L.) in the northern extreme of the Mississippi Embayment (Figure 8). The wetland is highly productive, with a continual coat of duckweed on the water surface, and is estimated to be flooded by the adjacent Cache River almost every year as illustrated in Figure 2. In the case of this riparian wetland, approximately 447 g m^{-2} of sediments and about 3.6 g m^{-2} of phosphorus was retained by the wetland during a flood event (Table 1). Gross sedimentation rates were estimated to be 5,600 g $m^{-2}y^{-1}$ from sediment traps, with most of that due to allochthonous productivity and resuspension. The flux of phosphorus with flooding represented the largest inflow of the nutrient during the study period. Our preliminary phosphorus budget for the system suggested that the flooding river deposited about 33 times as much phosphorous as entered the wetland through bulk precipitation and that the 30 ha swamp retained about 0.4 percent of the total annual phosphorus flux of the river. For the year in which measurements took place, the swamp retained about 10 times more phosphorus from the flooding river than it returned by surface and groundwater to the river for the remainder of the year (Mitsch et al., 1979).

Old Woman Creek, Ohio - A Natural Instream Wetland in a Coastline Environment

Old Woman Creek State Nature Preserve and National Estuarine Research Reserve is a coastal wetland located adjacent to Lake Erie in Erie County, Ohio (Figure 9). The wetland itself is 30 ha and extends about 1 km south of the Lake Erie shoreline (see Mitsch 1988, 1989; Klarer and Millie, 1989; Mitsch et al, 1989; Reeder, 1990; Mitsch and Reeder, 1991a,b) and is a good example of an instream natural wetland such as the one shown in Figure 1. It is approximately 0.34 km wide at its widest point, wetland depths may reach up to 3.6 m in the inlet stream channel but are usually less than 0.5 m. Klarer and Millie (1989) estimated that the retention time of the wetland varies between 24 hours (at peak flow) and 114 hours (at average flow). The wetland has an outlet to Lake Erie that is often open but that can be closed for extended periods by shifting sands forming a barrier beach. Rare but dramatic seiches on Lake Erie can reverse the flow, causing lake water to spill into the wetland. Most of the aquatic habitats within the wetland are open-water planktonic systems with about 30 percent of the wetland as floating leaved beds of American lotus (*Nelumbo luteau*). The major land use within the watershed (68.6 km^2) is agricultural, and hence the wetland is a receiving system for significant loadings of NPS pollution, primarily in the spring. Sedimentation in the wetland was estimated to have been 0.76 mm y^{-1} prior to agricultural development in the early 1800s and more than 10 times that (10 mm y^{-1}) at present (Buchanan, 1982).

Table 2 contrasts rates of phosphorus retention in Old Woman Creek from empirical model calculations, from field data collected in 1988 (Reeder, 1990; Mitsch and Reeder 1991a), and from a simulation model developed for the system (Mitsch and Reeder, 1991b). The empirical phosphorus retention model of Richardson and Nichols (1985) predicted 8-19 mg-P $m^{-2}d^{-1}$ being permanently buried in the wetland sediments under standard phosphorus loading rates from Lake Erie watersheds. This method of estimating phosphorus retention does not take into account any of the interactions between Lake Erie and the wetland, nor does it consider any intrasystem interactions.

A mass balance predicted from field data (Reeder, 1990; Mitsch and Reeder, 1991a) has a much lower loading rate, primarily because the data reflect drought year conditions and do not include an entire calendar year. Phosphorus retention from the field data was estimated to be 36 percent of the inflow, nearly the same retention percentage as predicted by the empirical models. The field data results indicate that the contribution of Lake Erie to phosphorus loading to the wetland is minimal (0.4 percent of watershed inflow). Actual loading rates and retention may be higher than either the empirical model or field data predict because data from one recent sediment core show that an average of 22 mg-P $m^{-2}d^{-1}$ has been deposited over the past 180 years (Reeder, 1990). However, the sediment coring site is in the area of the marsh with the highest sedimentation rate (Buchanan, 1982), so the sedimentation data probably represent a region of maximum retention.

A simulation model developed for Old Woman Creek takes into account the effects of hydrology on loading rate, the amount of transformation occurring due to the biota, and the sedimentation and resuspension estimated from model calibration. The simulation model predicts that 2.9 mg-P $m^{-2}d^{-1}$ or less than 10 percent, of the phosphorus inflow is permanently retained in the wetland-more than the field estimates suggest but less than the empirical model and sediment core predict. As expected the primary producers transform some bioavailable phosphorus into nonbioavailable forms, but this contribution to the sediments is minimal (2.2 percent) in the model. Other simulations predicted that as much as 12.4 mg-P $m^{-2}d^{-1}$ could be retained in the marsh, especially under high water-level conditions (Reeder, 1990; Mitsch and Reeder, 1991b). The model estimates are reasonable considering the variability of the data used for comparison, and probably represent the best available estimate on phosphorus retention in this Great Lakes coastal wetland.

The Des Plaines River Wetlands, Illinois – Experimenting with Watershed Size

The restoration of entire rivers has been shown to be an elusive goal in many parts of the world because of years of drainage and channel "improvement," floodplain development, and significant loads of sediments and other nonpoint pollutants. Even in perusing pollution control, we have paid too much attention to the stream itself and not enough to its interactions with its floodplain.

One ecological engineering project on the Des Plaines River north of Chicago in Lake County, Illinois, involves restoration of a length of a river floodplain and establishing experimental wetland basins (Figure 10). This project, begun in 1982 as the "Des Plaines River Wetland Demonstration Project," has as its goals "to demonstrate how wetlands can benefit society both environmentally and economically, and to establish design procedures, construction techniques, and management programs for restored wetlands" in a riverine setting (Hey et al., 1989). Four experimental wetland basins (1.9 to 3.4 ha) have been constructed and instrumented at the northern half of the site for precise hydrologic control to achieve ecosystem experimentation conditions for investigating hydrologic design of wetlands subjected to inputs of NPS pollutants (Table 3). Two of the wetlands had inflows of 35 to 38 cm/wk (high flow) while the other two wetlands had inflows of 10 to 16 cm/wk (low flow). In comparison, precipitation contributed less than 2 cm/wk during this first year of experimentation. Water level was maintained at approximately the same depth in each wetland basin.

Results from the experimental wetlands after 1 year of study point to substantial productivity in all of the wetlands with most of the phosphorus and sediments from the river retained by the wetlands. Tables 4 and 5 illustrate sediment and phosphorus budgets for the four experimental wetlands. The low flow wetlands retained from 93 to 98 percent of the inflow of sediments while sediment retention by the high flow wetlands was 88 to 90 percent (Table 4).

Table 4 also shows estimates of macrophyte and plankton production for the wetlands, and sedimentation rates as measured by sediment traps. Sedimentation as measured by the traps was much greater than that estimated by subtracting outflow sediment from inflow sediment. Some of this difference is undoubtedly due to autochthonous productivity (macrophyte plus plankton production), although these productivity estimates are small relative to sedimentation rates measured by, the traps. Overall, the results shown in Table 4 suggest that 1) the sedimentation traps include a significant amount of resuspension from the wetland sediments and 2) the autochthonous production of sediments (mostly organic) is generally greater than the accumulation of sediments due to the pumped water.

The phosphorus budget for the wetlands is shown in Table 5. Resuspension of sediments is obviously an important source of phosphorus to the sediment traps. Of greater interest, the calculated uptake of phosphorus by, the macrophytes and water column algae and aquatic plants is similar to net retention as estimated by input minus output calculations. This illustrates that these wetlands are probably not being overloaded by phosphorus and may, in fact, be close to a sustainable loading of phosphorus.

Comparison

A comparison of natural and constructed wetlands for the control of NPS loadings is enlightening (Table 6). The natural wetlands discussed represent both a riparian and an instream wetland. Heron Pond is flooded by approximately 10 percent of the floodwater from a 1,570 km^2 watershed and had approximately 15,000 g $m^{-2}y^{-1}$ of sediments pass over it during flooding, while an estimated 500 g m 2y-1.ere retained. By contrast, the Des Plaines River wetlands were subjected to inflows of 200 to 900 g m 2y-1 of sediments and retained from 200 to 800 g $m^{-2}y^{-1}$ In the natural wetland, 3 percent of the sediments are retained; in the constructed riparian wetlands with pumps, about 90 percent of the sediments are retained.

The comparison of the wetland types for phosphorus retention is similar. In Heron Pond, the riparian wetland retained about 4.5 percent of the phosphorus that passed over it during the flood event. At Old Woman Creek, an instream wetland fed by a 69 km^2 .atershed retained approximately, 10 percent of the phosphorus that entered. Retention of phosphorus by the constructed Des Planes River wetlands was much higher, with 63-68 percent of the phosphorus retained in the high-flow wetlands and 83-96 percent retained in the low flow wetlands. Phosphorus retention as a function of phosphorus loading illustrates that wetlands built in downstream locations will be generally less efficient in retaining nutrients than would smaller upstream wetlands. But looking at efficiency alone would be misleading; wetlands built downstream could retain more mass of nutrients. A tradeoff is obviously necessary.

CONCLUSIONS

We must tie our rivers and wetlands back together in an ecologically sustainable way. Successful ecological engineering requires that we take advantage of our increasing knowledge of ecology and its principles (e.g., succession, energy flow, self-design, etc.) to construct and restore wetlands as part of a natural landscape with minimum human maintenance. But we must resist the ever-present temptation to over-engineer by channeling energies that cannot be channeled and to over-biologize by introducing species that the design does not support. Boule (1988) has recommended that the human contribution to the design of wetlands be kept simple without reliance on complex technological approaches that invite failure. He states that "simple systems tend to be self-regulating and self-maintaining." Self-design and self-organization are in fact the very cornerstones of the ecological engineering approach and should be used whenever possible in designing wetlands in the landscape.

ACKNOWLEDGEMENTS

The author appreciates the thoughtful and complete reviews provided by Mark M. Brinson, Robert L. Knight, and Richard Olson and the editing assistance of Ruthmarie H. Mitsch.

REFERENCES

Boule, M. E., 1988. Wetland creation and enhancement in the Pacific Northwest. pp. 130-136. In: J. Zelazny and J. S. Feierabend (eds.), Proceedings of the Conference on Wetlands: Increasing Our Wetland Resources, Washington, DC. Corporate Conservation Council, National Wildlife Federation, Washington, DC.

Brooks, R. P., 1989. Wetland and waterbody restoration and creation associated with mining. pp. 529-548. In: J. A. Kusler and M. E. Kentula (eds.), Wetland Creation and Restoration: The Status of the Science. Island Press, Washington, DC.

Buchanan, D., 1982. Transport and deposition of sediment in Old Woman Creek Estuary of Lake Erie. M.S. Thesis, The Ohio State University, Columbus, OH.

Dahl, T. E. 1990. Wetlands losses in the United States 1780's to 1980's. U.S. Department of Interior, Fish and Wildlife Service, Washington, DC.

Davis, D. G., 1989. No net loss of the nation's wetlands: a goal and a challenge. Water Environment and Technology, 4: 513-514.

Hey, D. L., M. A. Cardamone, J. H. Sather, and W. J. Mitsch, 1989. Restoration of riverine wetlands: The Des Plaines River wetlands demonstration project. pp. 159-183. In: W. J. Mitsch and S. E. Jorgensen (eds.), Ecological Engineering: An Introduction to Ecotechnology. John Wiley & Sons, New York, NY.

Klarer, D. M. and D. F. Millie, 1989. Amelioration of storm-water quality, by a freshwater estuary. Archiv fur Hydrobiologia, 116: 375-389.

Livingston, E. H., 1989. Use of wetlands for urban stormwater management. pp. 253-264. In: D. A. Hammer (ed.), Constructed Wetlands for wastewater Treatment. Lewis Publishers, Inc., Chelsea, MI.

Loucks, 0. L., 1989. Restoration of the pulse control function of wetlands and its relationship to water quality objectives. pp. 467-478. In: J. A. Kusler and M. E. Kentula (eds.), Wetland Creation and Restoration: The Status of the Science. Island Press, Washington, DC.

Mitsch, W. J. (ed.), 1988. Ecological engineering and ecotechnology with wetlands: applications of systems approaches. pp. 565-580. In: A. Marani (ed.), Advances in Environmental Modelling. Elsevier, Amsterdam.

Mitsch, W. J. (ed.), 1989. Wetlands of Ohio's coastal Lake Erie: a hierarchy of systems. Final Report, Ohio Sea Grant College Program, Columbus, OH.

Mitsch, W. J., 1990. Wetlands for the control of nonpoint source pollution: preliminary feasibility study for Swan Creek watershed of northwestern Ohio. Final Report to Ohio Environmental Protection Agency, The Ohio State University, Columbus, OH.

Mitsch, W. J., 1991. Ecological engineering: approaches to sustainability, and biodiversity in the U.S. and China. pp. 428-448. In: R. Costanza (ed.), Ecological Economics: The Science and Management of Sustainability. Columbia University Press, New York, NY.

Mitsch, W. J. and J. K. Cronk. Creation and restoration of wetlands: some design considerations for ecological engineering. Advances in Soil Science: in press.

Mitsch, N. J., C. L. Dorge, and J. R. Wiemhoff, 1977. Forested Wetlands for Water Resource Management in Southern Illinois. Research Report No. 132, Illinois Water Resources Center, University of Illinois, Urbana, IL.

Mitsch, W. J., C. L. Dorge, and J. R. Wiemhoff, 1979. Ecosystem dynamics and a phosphorus budget of an alluvial cypress swamp in southern Illinois. Ecology, 60: 1116-1124.

Mitsch, W. J. and S. E. Jorgensen, 1989. Introduction to ecological engineering. pp. 3-12. In: W. J. Mitsch and S. E. Jorgensen (eds.), Ecological Engineering: An Introduction to Ecotechnology. John Wiley & Sons, New York.

Mitsch, W. J., and B. C. Reeder, 1991a. Nutrient and hydrologic budgets of a Great Lakes coastal freshwater wetland during a drought year. Wetlands Ecology and Management, 1(3): in press.

Mitsch, W. J. and B. C. Reeder, 1991b. Modelling nutrient retention of a freshwater coastal wetland: estimating the roles of primary productivity, sedimentation, resuspension and hydrology. Ecological modelling, 54: 151-187.

Mitsch, W. J., B. C. Reeder, and D. M. Klarer, 1989. The role of wetlands for the control of nutrients with a case study of western Lake Erie. pp. 129-159. In: W. J. Mitsch and S. E. Jorgensen (eds.), Ecological Engineering: An Introduction to Ecotechnology. John Wiley & Sons, New York, NY.

National Wetlands Policy Forum, 1988. Protecting America's Wetlands: an Action Agenda. The Conservation Foundation, Washington, DC.

Newton, R. B., 1989. The effects of stormwater surface runoff on freshwater wetlands: a review of the literature and annotated bibliography. University of Massachusetts, Amherst, MA.

Odum, H. T., 1989. Ecological engineering and self-organization. pp. 79-101. In: W. J. Mitsch and S. E. Jorgensen (eds.), Ecological Engineering: An Introduction to Ecotechnology. John Wiley & Sons, New York, NY.

Ogawa, H. and J. W. Male, 1983. The flood mitigation potential of inland wetlands. Water Resources Research Center Publication No. 138, University of Massachusetts, Amherst, MA.

Reeder, B. C., 1990. Primary productivity phosphorus cycling, and sedimentation in a Lake Erie coastal wetland. Ph.D. dissertation, The Ohio State University, Columbus, OH.

Richardson, C. J. and D. S. Nichols, 1985. Ecological analysis of wastewater management criteria in w'etland ecosystems. pp. 351-391. In: P. J. Godfrey, E. R. Kaynor, S. Pelczarski, and J. Benforado (eds.), Ecological Considerations in Wetlands Treatment of Municipal wastewaters. Van Nostrand Reinhold Company, New York, NY.

Table 1. Sediment and Phosphorus Budgets for Heron Pond Riparian Cypress Swamp, Adjacent to Cache River, Southern Illinois (From Mitsch et al., 1979).

	River Load*	Passes over Wetland*	Retained by Wetland*	Percent Retained
Sediments	–	15,000	447	3
Total Phosphorus	880	80.2	3.6	4.5

* in g $m^{-2}y^{-1}$.

Table 2. Comparison of Nutrient Retention Capabilities of Old Woman Creek Wetland by Different Measurements (measurements are in mg-P $m^{-2}d^{-1}$).

Method	Inflow	Outflow	Retention	Source
Empirical Model v. I	33-63	–	14-19	Mitsch et al., 1989
Empirical Model v. II	17-33	–	8-13	Mitsch and Reeder, 1991a
Field Data	2.2	1.4	0.8	Reeder, 1990; Mitsch and Reeder, 1991a
Sediment Core	–	–	22	Reeder, 1990
Simulation Model*	30.1	26.5	2.9	Mitsch and Reeder, 1991b

*for 9-month period only

Table 3. Hydrologic Experiment -- Des Plaines River Wetland Demonstration Project for Period October 1989 through September 1990.

	Wetland 3	Wetland 4	Wetland 5	Wetland 6
Size, acres	5.75	5.79	4.61	8.53
Size, hectares	2.33	2.34	1.87	3.45
Design Inflow, cm/wk	30	5	30	5
Inflow, cm/wk (n = 52)*	37.8 ± 2.2	9.8 ± 0.9	35.2 ± 2.2	16.1 ± 1.8
Outflow, cm/wk (n = 52)*	38.5 ± 1.9	9.0 ± 1.3	33.9 ± 2.0	4.1 ± 0.9
Precipitation, cm/wk	1.6			
Evapotranspiration, cm/wk	1.9			

*Numbers are average ± standard error

Table 4. Budget of Suspended Sediments for Des Plaines River Experimental Wetlands, October 1989 through September 1990 (units are in g-dry wt m^{-2} w^{-1}; inflows, outflows, sedimentation, and productivity are all estimated independently).

Experim. Wetland	Inflow	Outflow	Sedimentation				Macrophyte Production* (Growing Season)	Est. Plankton Production* (Growing Season)
			Winter (212 d)	Growing Season (151 d)	Wt. Average (363 d)	Increase in Summer		
3	17.4	2.1	117	381	227	264	18	13
4	4.1	0.3	13	353	154	340	24	11
5	16.6	1.7	25	199	97	174	26	18
6	4.2	0.1	15	178	83	163	6	16

*average of 1989 and 1990 estimated above-ground biomass accumulation divided by 22 weeks

**assumes aquatic efficiency for each wetland, average of 4,330 kcal m^{-2}-d^{-1}, and accumulation of 50 percent of gross primary productivity, averaged over summer week.

Table 5. **Budget of Total Phosphorus for Des Plaines River Experimental Wetlands, october 1989 through September 1990 (units are in mg-p m^{-2} wk^{-1}).**

Experim. Wetland	Inflow	Outflow	Sedimentation Traps Weighted Average*	Macrophyte Uptake**	Est. Algal Uptake***
3	45.1	16.5	186	22	4
4	10.3	1.7	126	32	3
5	43.7	14.0	91	23	5
6	13.7	0.5	85	5	5

*assumes approximately 1 mg-P g dry wt^{-1} in sediments in traps x annual sedimentation rate

**macrophyte biomass accumulation x weighted concentration, averages over 52 weeks

***assumes aquatic efficiency for each wetland, average 4,330 kcal m^{-2} d^{01}, 5-month growing season, accumulation of 50 percent of gross primary productivity, and 0.7 mg-P g dry wt^{-1} *(Cladophora)*, averaged over 52 weeks

Table 6. Comparison of Nonpoint Source Sediment and Phosphorus Retention in Natural and Constructed Wetlands

	watershed	sediments, g $m^{-2}y^{-1}$			phosphorus, g-P $m^{-2}y^{-1}$		
	sizw, km^2	loading	retention	(%)	loading	retention	(%)
Natural Wetlands							
Heron Pond, Southern IL (riparian)	1,570	15,000	447	(3)	80.2	3.6	(4.5)
Old Woman Creek, OH (instream)	69	---	---		11	1.1	(10)
Constructed Wetlands							
Des Plains River Wetlands (riparian with pump; year 1)							
Wetland 3 - high flow	1.4*	905	796	(88)	2.34	1.49	(63)
Wetland 4 - low flow	0.4*	213	198	(93)	0.54	0.45	(83)
Wetland 5 - high flow	1.0*	863	775	(90)	2.27	1.54	(68)
Wetland 6 - low flow	0.5*	218	213	(98)	0.71	0.69	(96)

*approximate equivalent watershet size for pumping rate

LIST OF FIGURES

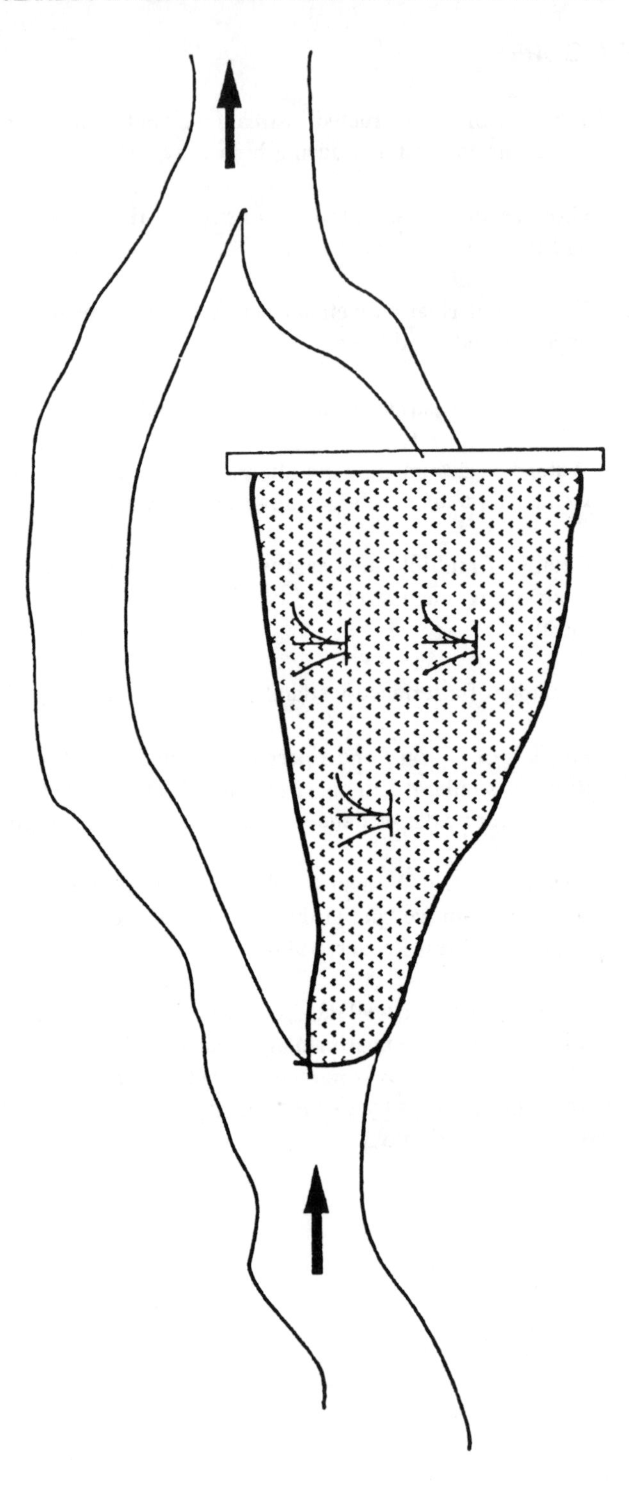

Figure 1

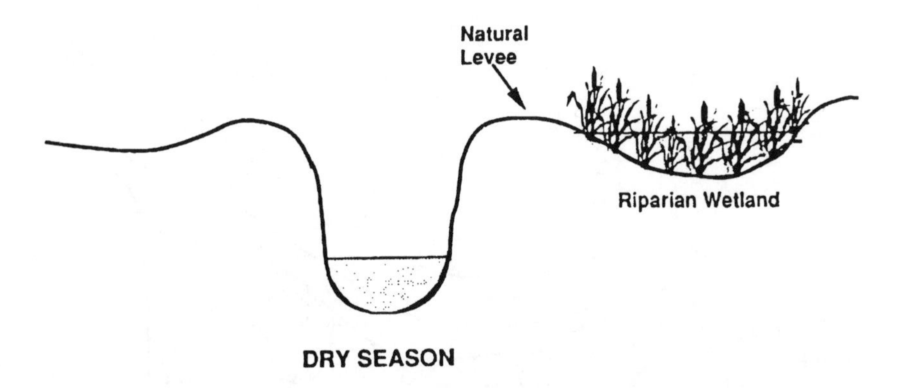
Natural
Levee
Riparian Wetland
DRY SEASON

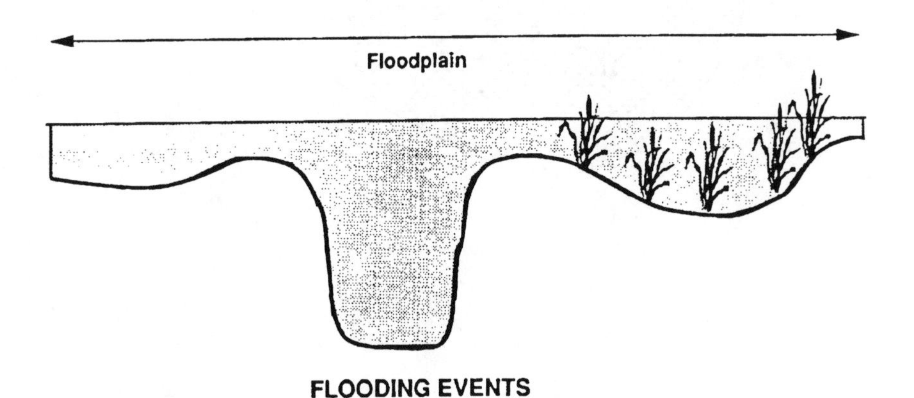
Floodplain
FLOODING EVENTS

Figure 2

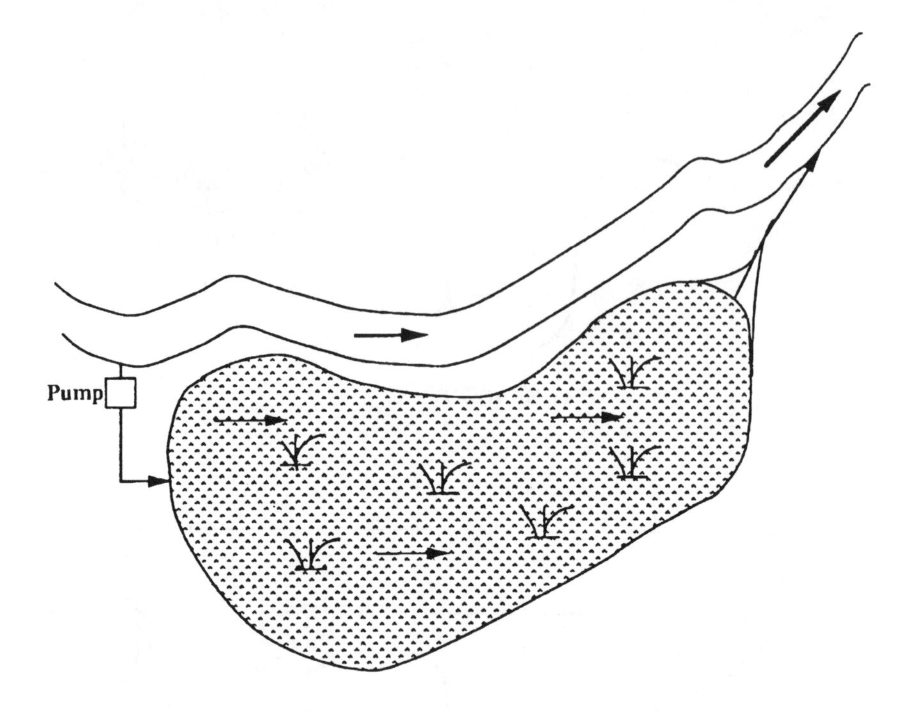

Figure 3

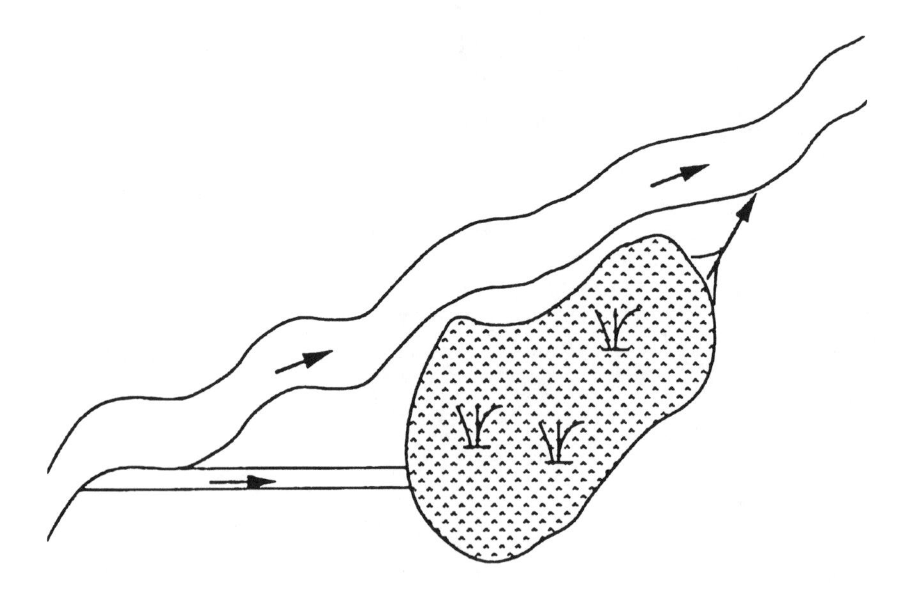

Figure 4

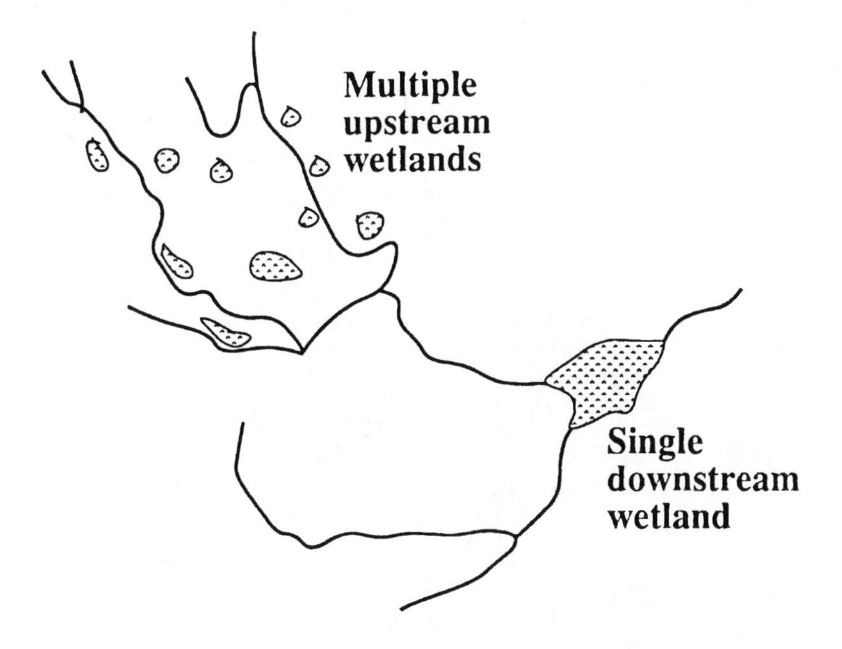

Figure 5

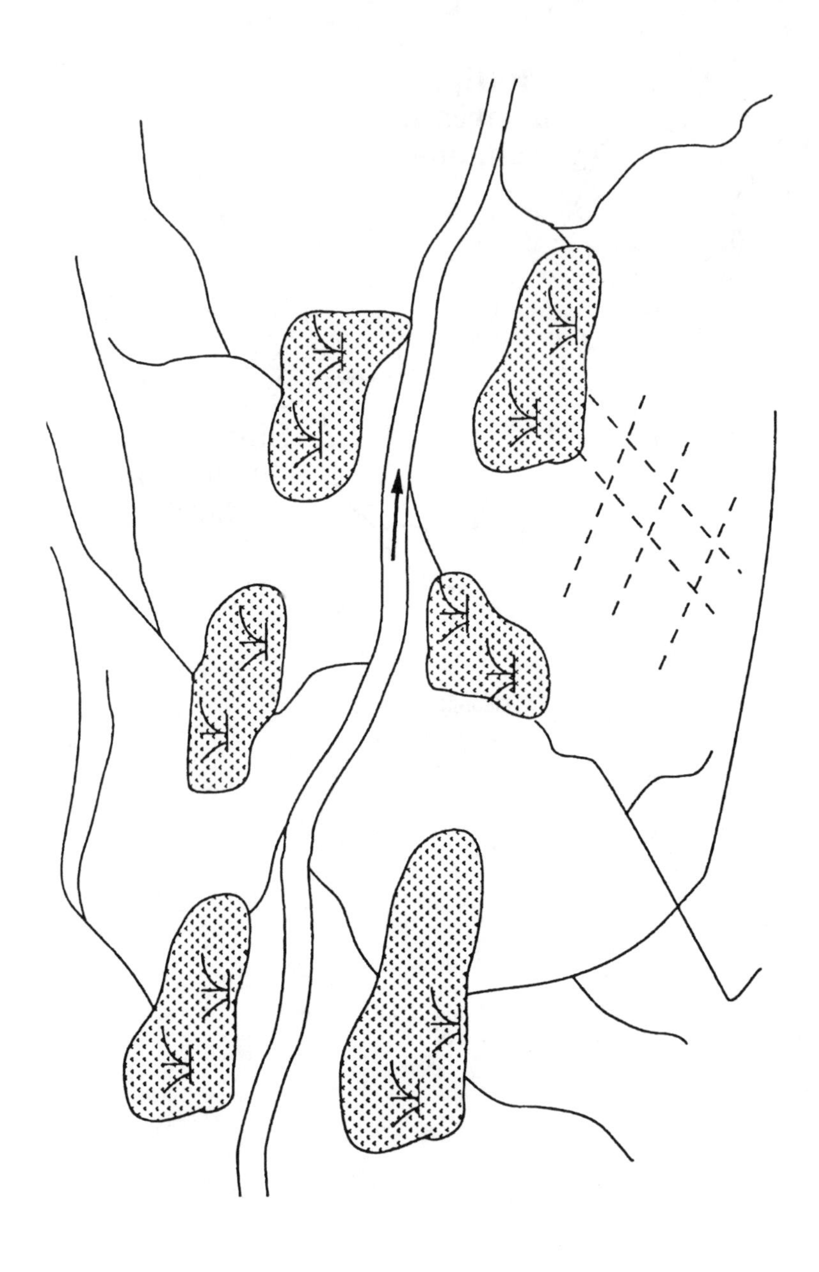

Figure 6

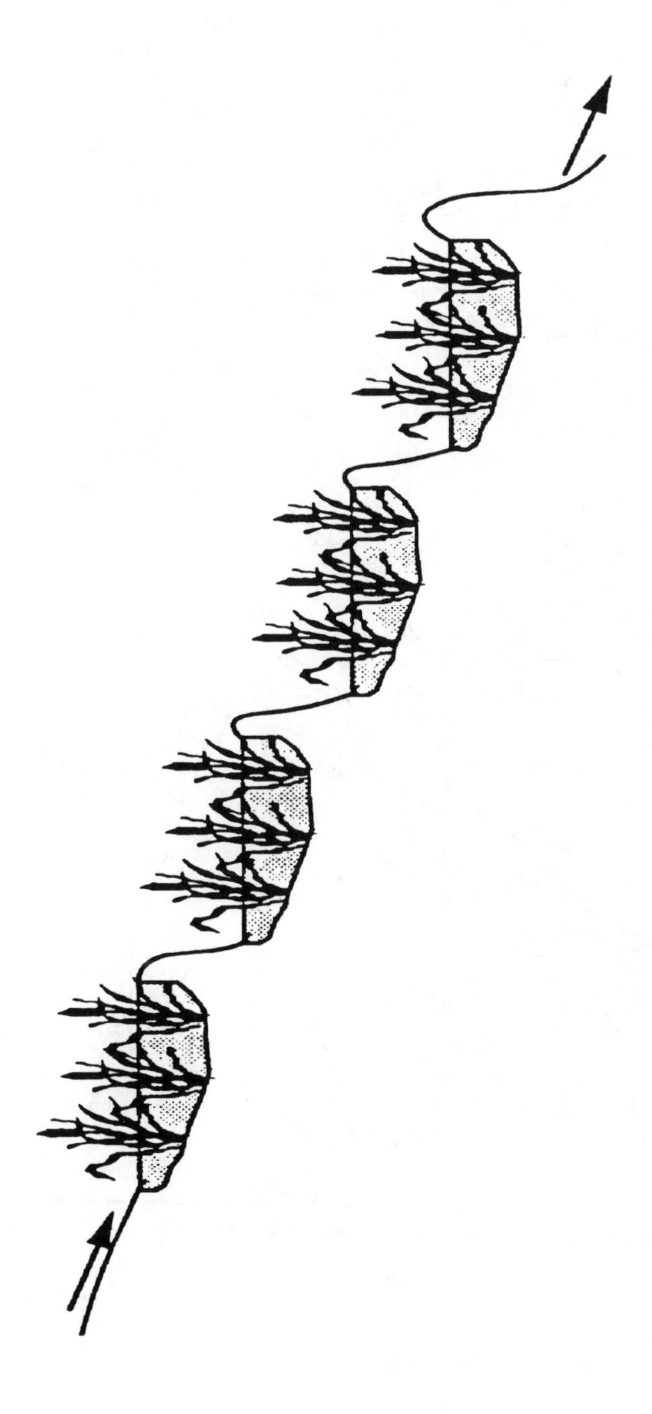

Figure 7

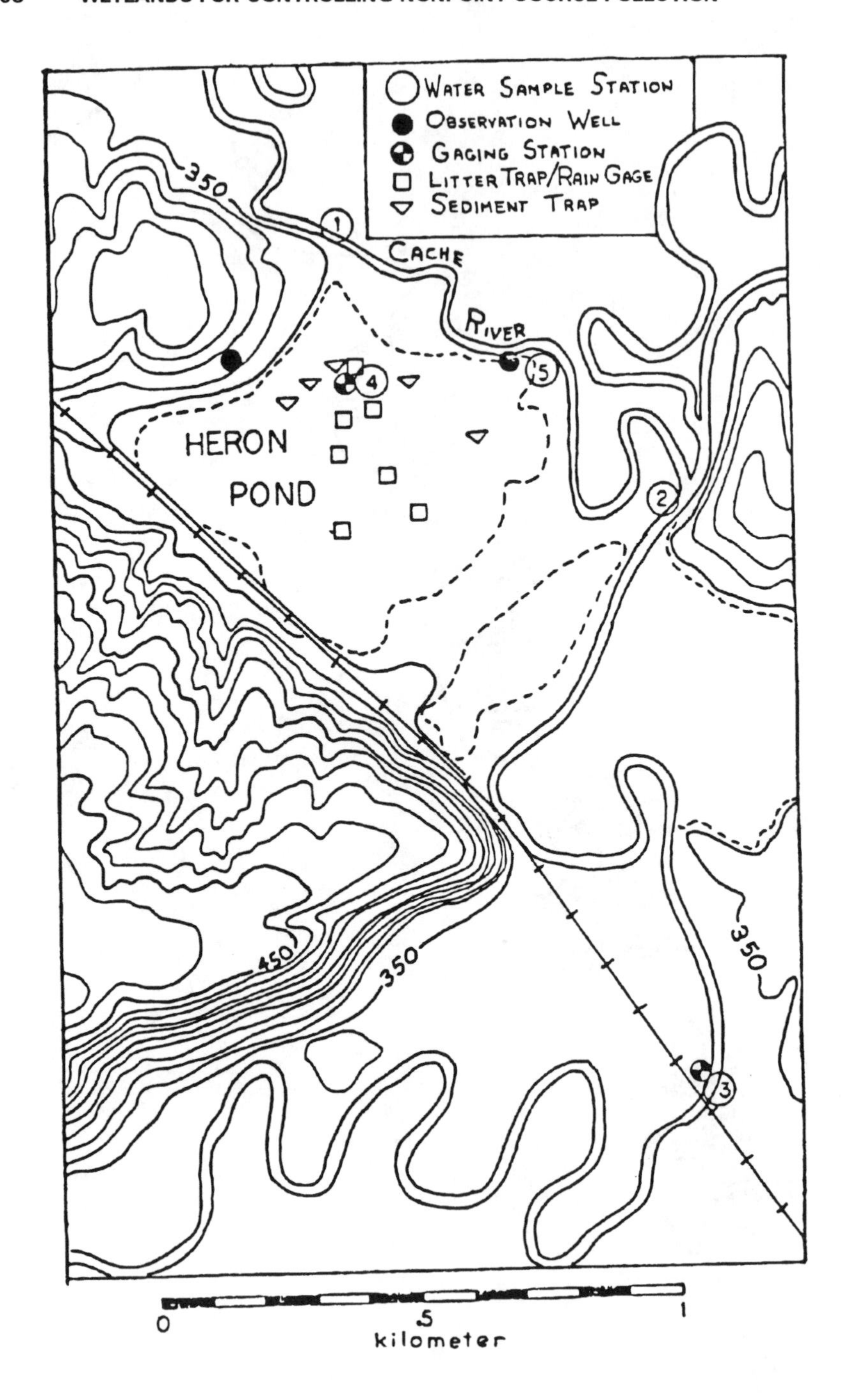
Water Sample Station
Observation Well
Gaging Station
Litter Trap/Rain Gage
Sediment Trap
350
Cache
River
HERON
POND
450
350
350
0
.5
1
kilometer

Figure 8

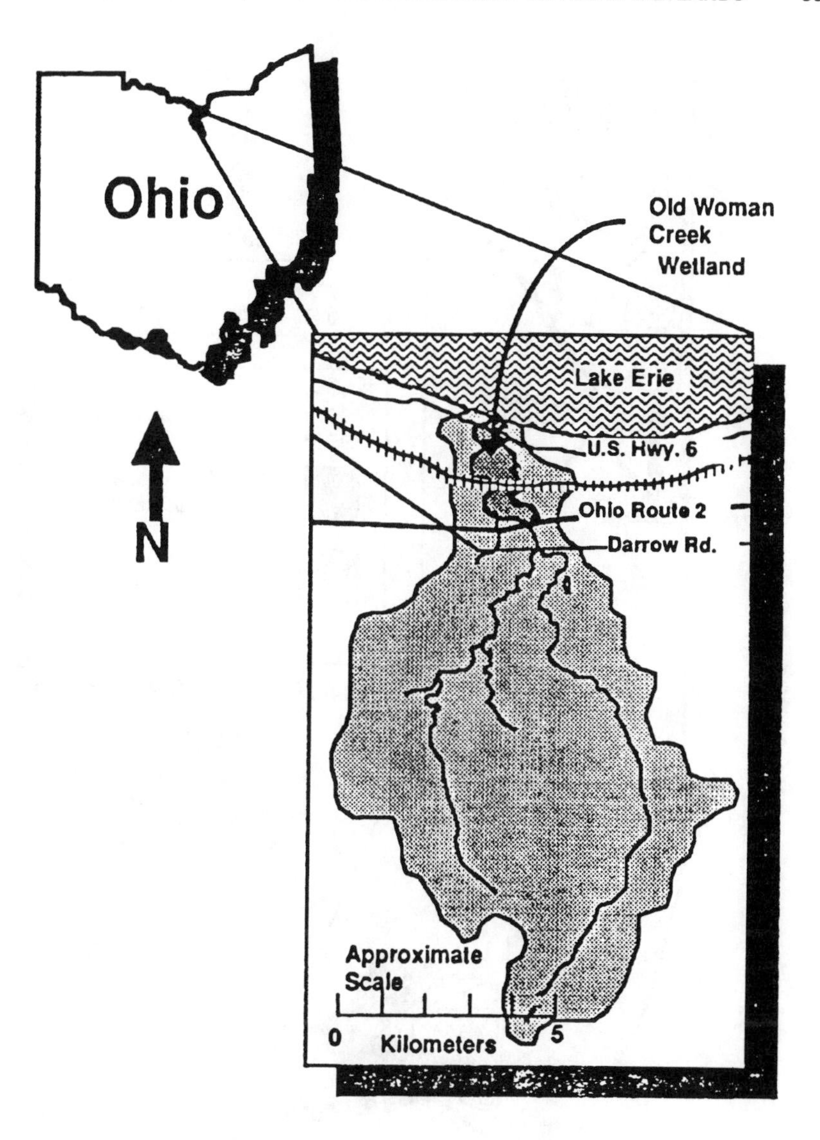
Ohio
Old Woman
Creek
Wetland
Lake Erie
U.S. Hwy. 6
Ohio Route 2
Darrow Rd.
N
Approximate
Scale
0
5
Kilometers

Figure 9

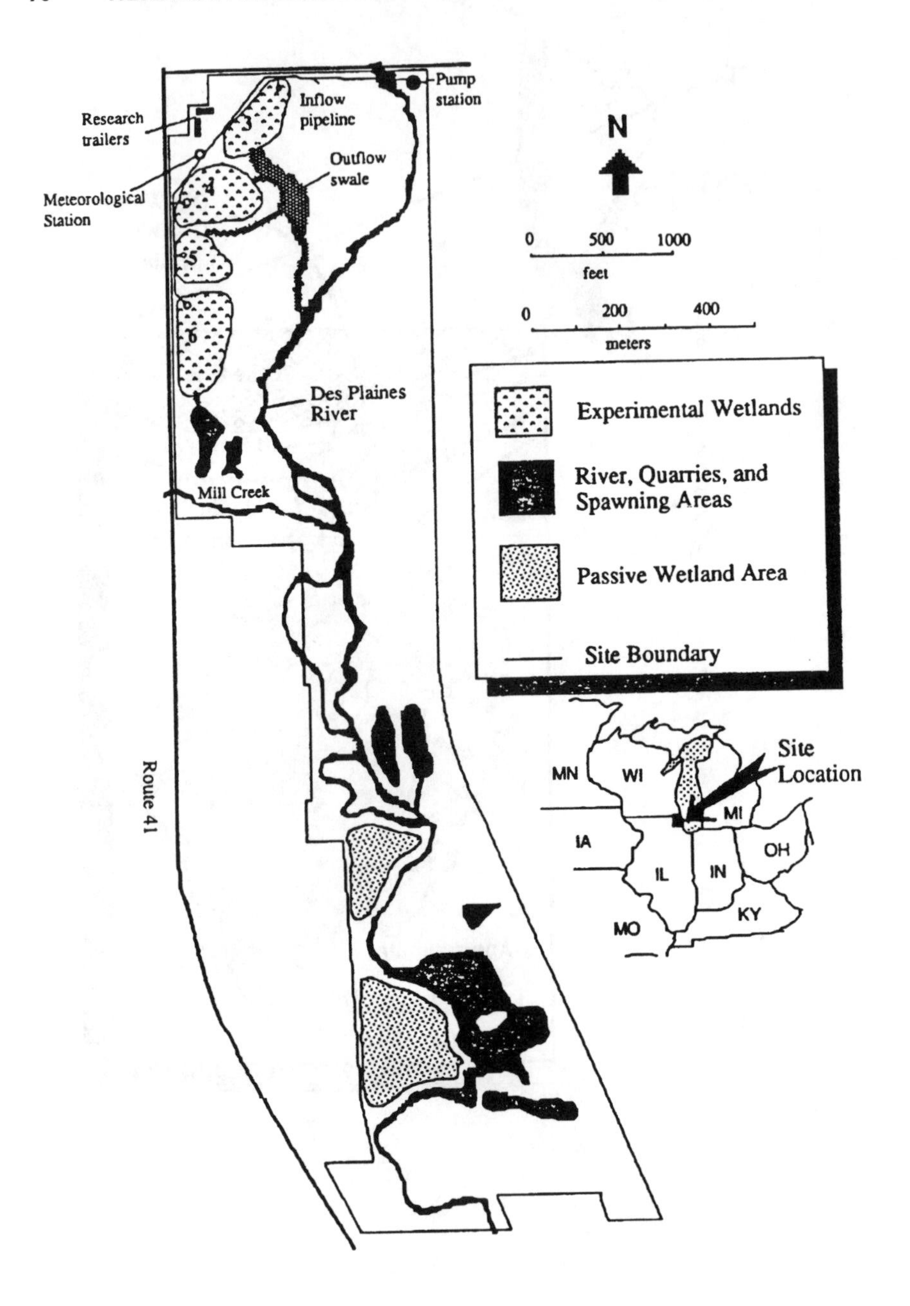

Pump station
Inflow pipeline
Research trailers
Outflow swale
Meteorological Station
N
0 500 1000
feet
0 200 400
meters
Des Plaines River
Experimental Wetlands
River, Quarries, and Spawning Areas
Mill Creek
Passive Wetland Area
Site Boundary
Site Location
Route 41
MN
WI
MI
IA
IL
IN
OH
MO
KY

Figure 10

CHAPTER 4

Designing Constructed Wetlands Systems to Treat Agricultural Nonpoint Source Pollution

Donald A. Hammer, Regional Waste Management Department, Tennessee Valley Authority

ABSTRACT

Increasingly concentrated animal husbandry practices and more intensive row crop farming have expanded agricultural pollution problems. Implementing accepted best management practices (BMPs) for erosion control and waste handling along with a combination of 1) onsite constructed wetlands, 2) nutrient-sediment control systems in small watersheds, and 3) natural wetlands along streams and at strategic locations in large watersheds may provide low-cost, efficient control. Design recommendations and examples are included.

INTRODUCTION

Public concern about water pollution during the last 30 years has resulted in State and Federal legislation regulating discharges and provided financial assistance for municipal treatment facilities. Substantial progress in treating point sources has and is continuing to occur, especially in larger cities and with major industries. Widespread implementation of wetlands treatment technology may accomplish similar objectives with small community and small industry sources. However, anticipated improvements in the Nation's waters have not been realized, and recent evaluations reflect a growing concern over nonpoint source (NPS) pollution, especially agricultural wastewater and agricultural cropland runoff, and urban stormwater runoff. These principal contributors to NPS pollution problems have been difficult to remedy with conventional wastewater treatment and soil/water conservation methods.

NPS pollution from agricultural operations, urban areas, failed home septic tank drain fields, mining, and other land disturbing activities continues to detrimentally impact 30 - 50 percent of our nation's waterways. Increasing regulatory focus on agricultural NPS pollution probably results not only from reduction or elimination of point sources but also from real changes in waste loading in receiving streams because of changing animal husbandry practices. Previously, free-ranging livestock (including poultry) at relatively low population densities caused little aquatic pollution because wastes were widely dispersed and natural soil systems recycled nutrients onsite. However, the historical and continuing tendency to confine livestock in ever smaller areas to improve production efficiency has the effect of concentrating animal waste loading with subsequent runoff to nearby streams. Concurrent removal of woody and non-woody riparian vegetation to increase efficiency by using all available acreage, or incidental removal due to livestock grazing and loafing has eliminated the buffer strip that formerly protected streams from direct pollutant impacts (Hammer, 1989a).

Farmers are unlikely to purchase and operate package treatment plants. Requiring a hog producer to purchase a multimillion dollar treatment system to deal with the waste from 1,000 hogs, which has similar organic loading but is much more concentrated than waste from a city of 1,000 residents, is unrealistic since many farmers are heavily in debt with marginally profitable operations. On the other hand, constructed wetlands waste treatment systems would seem more amenable to the substantial range of hydraulic and pollutant loading, temporal fluctuations, dispersed nature, and the need for low-cost, low-technology systems acceptable to farmers. Furthermore, planting and maintenance requirements on the farm differ little from skills needed in growing other crops, and land costs are relatively low.

NATURAL WETLANDS

Although natural wetlands systems provide many functional benefits to our society, virtually all can be grouped into three broad categories – life support, hydrologic buffering, and water quality improvement. Most of us are familiar with the many types and large numbers of animals, especially birds, that are dependent upon wetlands. But how many of us realize that the crayfish industry in Louisiana, the shellfish and much of the finish industry along our coasts, and the furrier industry that clothes our elegant women are also all dependent upon wetlands (Mitsch and Gosselink, 1986). Wetlands also reduce flooding along rivers and streams by reducing and desynchronizing peak runoff through slowing flood water velocities. At the other extreme, delayed flows emanating from wetlands augment base flows in streams and rivers maintaining levels essential for aquatic life. Finally, contaminated waters flowing through natural wetlands are cleansed by a combination of physical, chemical, and biological activities, and emerge as clean water (Hammer, 1990).

Wetlands ecosystems have intrinsic abilities to modify or trap a wide spectrum of water-borne substances commonly considered pollutants or contaminants. Doubtless, our ancestors perceived and exploited these abilities. Only in more recent times, however, have casual observations fostered renewed interest in wetlands. Such casual observations have led to investigations that documented changes in concentrations of various materials after processing by natural wetlands systems. In fact, much of the early work on constructed wetlands for wastewater treatment was stimulated by observing this purification phenomenon in natural wetlands systems (Seidel, 1971; Kadlec, et al., 1974; Odum and Brown, 1976; Small, 1977). For example, many observers have noticed accelerated soil erosion after heavy rains wash across unvegetated soils, and some have been fortunate enough to encounter situations were silt-laden waters transiting natural wetlands systems were readily compared with unprocessed waters. The striking visual differences were easily verified by sampling and analysis, and the information became an important component in a communal body of knowledge on natural wetlands values. Most ecologists believed this phenomenon was widespread, and a few even suggested that it might occur on a large scale, though little documentation was available.

I recently had the opportunity to observe an example of water quality improvements in river waters by a natural wetlands system on a very large scale in the Pantanal of western Brazil and adjacent portions of

Paraguay and Bolivia. This area is a large basin bordered by high plateaus and a savannah on the east, a semi-deciduous forest on the north, and a moderate mountain range and a semi-deciduous forest on the west. Runoff from these regions causes much of the 11,000,000 hectare area to be flooded from December to June, and a significant but unmeasured proportion is permanently wet. Many rivers enter the Pantanal from the eastern highlands, gradually disappearing and then reforming on the western and southern boundaries and draining off to the south. Over geological time, alluvial deposits of highlands silt has gradually transformed a flat or concave basin floor into a convex, dome-like surface with higher elevations in the center and lower on the margins.

Doubtless, this region provided important water improvement functions since tectonic forces created the original basin. But accelerated erosion and pollution from clearing and agricultural activities and other anthropogenic sources has tremendously increased the contaminant loading of rivers draining the plateaus on the east and north. The Rio Taquiri alone carries more than 30,000 metric tons of silt per day plus a variety of agrochemicals from soybean fields on the eastern plateau (Amaral, 1989). Other rivers transport lower silt loads, but most receive untreated sewage and industrial and mining pollution before reaching the Pantanal.

Amazingly, alarmingly high concentrations of silt and pollutants in inflowing river waters are reduced to innocuous levels in waters of rivers draining the region (Cadavid, 1989). Examination of a topographic map (abstracted in Figure 1) provides insight into the general process, though it does not reveal the complex of purification mechanisms. Notice the size, especially width, of the Rio Taquirj as it drops off the plateau and enters the Pantanal on the east. A fairly wide, deep, and fast flowing river courses out into the Pantanal, and rather quickly its width, depth, and velocity are reduced. A third of the way into the Pantanal, numerous small braided streams arise flowing perpendicularly out of the Rio Taquiri into the adjacent regions. Progressively increasing water loss with penetration into the Pantanal drastically reduces the Rio Taquiri until it almost disappears. In fact, the Pantanal functions as an 11,000,000 hectare sponge that absorbs inflowing waters, cleanses them of impurities, and slowly releases clean water through minor streams that aggregate into larger rivers along the southern and western boundaries. This large natural wetlands complex transforms heavily polluted influent waters into clean waters, and the slow release of waters collected during the rainy season augments base flow in the Rio Paraguay during the dry half of the year. On a large scale as well as in local areas, natural wetlands can perform substantial improvements in water quality and quantity.

Constructed wetlands have recently received considerable attention as low-cost, efficient means to clean up many types of wastewater. Though the concept of deliberately using wetlands for water purification has only developed within the last 20 years, in reality, human societies have indirectly used natural wetlands for waste management for thousands of years. We have always dumped our wastes into nearby streams or wetland areas. And as they do for natural terrestrial ecosystems, wetlands processed these wastes and discharged relatively clean water. However, as human populations increased and concentrated in towns and later cities, the increased quantity of wastes discharged into a small area soon overloaded natural wetlands and other aquatic systems damaging the wetlands and destroying their function in removing water borne pollutants. Without wetlands treatment buffering downstream areas, human wastes damaged aquatic life in rivers, bays, and oceans and threatened drinking water supplies.

WETLANDS PURIFICATION FUNCTIONS

Vegetation

Water purification functions of wetlands are dependent upon four principle components—vegetation, water column, substrates, and microbial populations. The principle function of vegetation in wetlands systems is to create additional environments for microbial populations (Pullin and Hammer, 1989). Not only do stems and leaves in the water column obstruct flow and facilitate sedimentation, they also provide substantial quantities of surface area for attachment of microbes and constitute thin-film reactive surfaces. In addition to the microbial environments in the water column of lagoons, wetlands have much additional surface area on portions of plants within the water column. Plants also increase the amount of aerobic microbial environment in the substrate incidental to the unique adaptation that allows wetlands plants to thrive in saturated soils. Most plants are unable to survive in water-logged soils because their roots cannot obtain oxygen in the anaerobic conditions rapidly created after inundation. However, hydrophytic, or wet-growing plants, have specialized structures somewhat analogous to a mass of breathing tubes in their leaves, stems, and roots that conduct atmospheric gases, including oxygen, down into the roots.

Because the outer covering on the root hairs is not a perfect seal, oxygen leaks out creating a thin film aerobic region – the rhizosphere – around each and every root hair. The larger region outside the rhizosphere remains anaerobic but the juxtaposition of a large, in-aggregate, thin film aerobic region surrounded by an anaerobic region is crucial to transformations of nitrogenous compounds and other substances. Wetlands vegetation substantially increases the amount of aerobic environment available for microbial populations, both above and below the surface. Wetlands plants generally take up only very small quantities (<5 percent) of the nutrients or other substances removed from the influent waters. However, some systems incorporating periodic plant harvesting have slightly increased direct plant removals at considerable operating expense.

Microbial Organisms

Microbes-bacteria, fungi, algae, and protozoa-alter contaminant substances to obtain nutrients or energy to carry out their life cycles. In addition, many naturally occurring microbial groups are predatory and will forage on pathogenic organisms. The effectiveness of wetlands in water purification is dependent on developing and maintaining optimal environments for desirable microbial populations. Fortunately, these microbes are ubiquitous, naturally occurring in most waters , and likely to have large populations in wetlands and contaminated waters with nutrient or energy sources. Only rarely, with very unusual pollutants, will inoculation of a specific type or strain of microbes be needed.

Substrates

Substrates various soils, sand, or gravel provide physical support for plants; reactive surface area for complexing ions, anions, and some compounds; and attachment surfaces for microbial populations. The water column – surface and subsurface water – transports substances and gases to microbial populations, carries off byproducts, and provides the environment and water for biochemical processes of plants and microbes.

Animals

Invertebrate and vertebrate animals harvest nutrients and energy by feeding on microbes and macrophytic vegetation, recycling and in some cases transporting substances outside the wetlands system. Functionally, these components have limited roles in pollutant transformations, but they often provide substantial ancillary benefits (recreation/education) in successful systems. In addition, vertebrate and invertebrate animals serve as highly visible indicators of the health and well-being of a marsh ecosystem, providing the first signs of system malfunction to a trained observer. Some invertebrates and many vertebrates occupy upper trophic levels within the system that are dependent upon robust, healthy populations of micro and macroscopic organisms in the critical lower levels. Declines in lower-level populations (including those involved in pollutant transformations) are reflected in changes in more visible animals in the higher levels. However, observations on types and numbers of indicator species must be carefully interpreted and/or compared to conditions in natural, unimpacted wetlands by an experienced wetlands ecologist, since certain species thrive in overloaded, poorly operating systems.

CONSTRUCTED WETLANDS

Constructed wetlands, in contrast to natural wetlands, are man-made systems that are designed, built, and operated to emulate natural wetlands or functions of natural wetlands for human desires and needs. As used for wastewater treatment, constructed wetlands may include swamps—wet regions dominated by trees, shrubs, and other woody vegetation—or bogs, which are low-nutrient, acidic waters dominated by *Sphagnum* or other mosses. However, we most commonly refer to them as marshes. Marshes are shallow-water regions dominated by emergent herbaceous vegetation—cattails, bulrushes, rushes, and reeds—and are adapted to a tremendous variety of soil and climatic conditions. Some marsh plants occur on every continent except Antarctica. Marshes are also adapted to a wide range of water quality conditions as well as substantial fluctuations in water flows and depths. Although bogs and swamps have been used for wastewater treatment, both are difficult to establish or manage and require fairly stable water quality and quantity conditions. Alterations in either are likely to cause undesirable changes in the structure and function of bogs and swamps.

The vast majority of wetlands constructed for wastewater treatment are classified as surface flow or free-water surface systems (Reed, 1990), that is, influent waters that flow across and largely above the surface of the substrate materials. Substrates are generally native clay or soil. In the other major class, subsurface-flow systems, waters flowing through the system pass entirely within the substrate, and free water is not visible. Substrates in subsurface-flow systems are typically various sizes of gravel or crushed rock. Though subsurface-flow systems appeared to have considerable potential only a few years ago, in practice virtually all subsurface-flow systems with 2 or more years of operating history have experienced serious clogging problems. In addition subsurface flow systems are unable to maintain adequate dissolved oxygen levels for ammonia removal. Thus, only a few subsurface-flow systems that treat municipal waste are operating in North America, but many of the European municipal systems are of this type. Only surface-flow systems have been used for mine drainage, agricultural waste, urban stormwater, industrial wastewaters, or other applications to date. Because a number of the operational subsurface flow systems have experienced clogging problems, only surface flow systems can be recommended for anything less than tertiary polishing of effluent with low concentrations of nutrients.

Constructed wetlands currently treat wastewaters from towns and small cities, mine drainage, urban stormwater runoff, livestock production, failed septic tank fields, land fill leachate, paper mills, tanneries, food processing plants, petroleum refineries, and many other small industrial sources (Reddy and Smith, 1987). Operating systems are located from sea level to 5,000' and from the tropics to subartic regions in Ontario and the Scandinavian countries (Miller, 1989; Brix and Schierup, 1989). Since operation is dependent on chemical and biological processes, pollutant removal efficiencies decline somewhat during low temperatures but discharge levels remain well below permit limits.

NONPOINT SOURCE POLLUTION

The most efficient approach to controlling agriculturally related NPS pollution—and the most acceptable to landowners—employs a combination of accepted BMP for waste handling and erosion control along with constructed and natural or restored wetlands systems in a hierarchial system. Normal BMP's include dry-stacking and disposal of solid wastes, roof guttering, lagoons, land application, terraces, grassed waterways, nutrient management, conservation tillage, and crop rotations. Following installation and/or use of BMPs, first-order control uses constructed wetlands designed and operated specifically for treating wastewater emanating from concentrated livestock areas, processing facilities, and in many cases, septic tanks serving the farm household.

Second-order control consists of nutrient/sediment treatment systems strategically located downstream from the wetlands treatment systems, at the lower end of grassed waterways and within intermittent stream courses throughout the individual farm (Figure 2). Third-order control deploys nutrient/sediment treatment systems, constructed wetlands/pond complexes, and restored or created wetlands at specific sites within a watershed that may include many individual farms. Fourth order systems consist of larger wetlands in the lower reaches of an individual watershed that function primarily for hydrologic buffering, and life support values in addition to limited water purification (Figure 3).

From a different perspective, first-order control may be considered simply wastewater treatment at the source, second order is treatment with constructed systems of less concentrated, aggregate wastewater from a variety of sources, third order is represented by buffer strips of riparian wetlands along permanent streams, small restored or created wetlands at specific points in the upper reaches of the watershed, and fourth order consists of larger areas of restored or created wetlands at tributary stream intersections in the lower sections of the watershed. Furthermore, first-order systems are principally designed and operated for wastewater treatment; second-order systems provide treatment but also produce some ancillary benefits; while third- and fourth-order systems function much the same as regional, natural wetlands accomplishing water purification, hydrologic (flood) buffering, life support, and related beneficial values.

Unfortunately, a few of the wetlands descriptors have been used synonymously and need precise definition to insure common understanding.

Natural wetlands are those areas wherein, at least periodically, the land supports predominantly hydrophytes and the substrate is predominantly untrained hydric soil or the substrate is non-soil and is saturated with water or covered by shallow water at some time during the growing season of each year. Natural wetlands have and continue to support wetlands flora and fauna.

Restored wetlands are areas that previously supported a natural wetlands ecosystem but were modified or changed, eliminating typical flora and fauna, and used for other purposes. These areas have then subsequently been altered to return to poorly drained soils and wetlands flora and fauna to enhance life support, flood control, recreational, educational, or other functional values.

Created wetlands formerly had well-drained soils supporting terrestrial flora and fauna but have been deliberately modified to establish the requisite hydrological conditions producing poorly drained soils and wetlands flora and fauna to enhance life support, flood control, recreational, educational, or other functional values.

Constructed wetlands consist of former terrestrial environments that have been modified to create poorly drained soils and wetlands flora and fauna for the primary purpose of contaminant or pollutant removal from wastewater. Constructed wetlands are essentially wastewater treatment systems and are designed and operated as such, though many systems do support other functional values.

Constructed and created wetlands occupy formerly dry sites; whereas, natural and restored systems occupy locales that historically were poorly drained and supported wetlands flora and fauna. In addition, wetlands built or managed for life support and other functional values would be subject to protective provisions with implementation of Section 404 regulations; whereas, wastewater treatment systems (constructed wetlands) requiring management practices that may be inimical to other functional values would be subject to NPDES provisions rather than 404 regulations.

In the holistic watershed approach discussed above, wetlands components of onsite wastewater treatment systems and sediment/nutrient control systems are constructed wetlands; whereas, riparian buffers, strategically placed marsh/pond complexes on permanent streams, and downstream marshes or bottomland hardwoods are created or restored wetlands. The former (first- and second-order) will require deliberate management/manipulation to maintain optimal treatment performance, but the latter (third- and fourth-order) would provide substantial water purification without active management and would support additional wetlands functions. First and second-order systems are located within the boundaries of an individual farm, and third order systems polish the runoff from a number of farms. In contrast, fourth-order systems control NPS pollution from an entire watershed.

Since most erosion control practices, solid waste handling, land application, lagoon treatment, riparian buffers, and large, fourth-order wetlands systems are fairly well understood and widely employed, the following discussion will be limited to methodologies of constructed wetlands for onsite treatment and the new nutrient/sediment control systems developed by Robert Wengryznek, Soil Conservation Service, Orono, Maine.

Livestock Wastewaters

In cooperation with the Soil Conservation Service (SCS) and Auburn University, the Tennessee Valley Authority (TVA), in 1988, initiated a constructed wetlands project to evaluate treatment performance and to develop design/operating criteria at Auburn's Sand Mountain Agricultural Experiment Station in northeast Alabama (Hammer, et al 1989). The system receives effluent from a secondary lagoon treating waste from approximately 500 hogs (Figure 4). The design includes a small farm pond for flood protection and dilution water, a mixing pond, and two replicates of five individual cells containing different vegetation. Different loading rates are applied to different cells to test plant survival and removal efficiencies under fairly extreme conditions (>100 mg/L NH^3. A similar experimental system treating wastewater from a dairy operation has recently been initiated by the SCS near Newton, Mississippi.

The farm pond was a less expensive alternative to excavating a flood water channel alongside the wetlands cell complex that serendipidously provided us an opportunity to demonstrate simple treatment for truly NPS runoff. Contoured terraces in the winter pasture above and alongside the farm pond direct virtually all of the runoff from the pasture to the upper end of the pond. High nutrient concentrations in pasture runoff supported an excessive algae bloom the first spring and summer after construction, and pond overflow discharging into a ditch flowing alongside the wetlands cell complex was of poor quality. However, merely installing a fence across the upper end and west side of the pond and planting wetlands vegetation on the pond margin and in the discharge channel has eliminated algae blooms and poor quality water in the ditch in subsequent years. The shallow, upper reaches of the pond are densely covered with *Fimbristylis, Carex, Juncus, and Scirpus*, which remove nutrients from the pasture runoff, eliminating impacts to the farm pond and downstream waters.

We also cooperated with Mississippi State University, the Mississippi Bureau of Pollution Control, and the SCS in constructing an operational constructed wetlands treatment system for a 500-hog operation at Mississippi State's Pontotoc Experiment Station (Figure 5). Since nitrogen (NH_2) is typically the most significant component of lagoon discharges, this design reflects current thinking on combining marsh-pond-marsh units within a single cell to improve nitrogen removal (Figure 6). I also added typical overland flow strips to compare removal efficiencies with the wetlands cells.

Rowcrop Runoff

In northern Maine, runoff from potato fields jeopardizes cold, deep lakes with a lake trout and land-locked salmon sport fishery that is economically significant to Aroostook County. In cooperation with other organizations, the Orono office of the SCS designed and constructed two demonstration nutrient/sediment control systems in watersheds with other BMPs for erosion control already in place. The nutrient/sediment system consists of a sedimentation ditch-bermed on the lower side-leading to an overland flow meadow, followed by a cattail marsh, a pond, and a final polishing meadow. Results have been very good with more than 80 percent removal of sediment, nitrogen, and phosphate from rowcrop runoff, and the treatment systems provide black duck breeding habitat as well as bait-fish rearing sites (Anon., 1991; Higgins, 1991).

DESIGNING A CONSTRUCTED WETLANDS FOR LIVESTOCK WASTEWATER TREATMENT

Constructed wetlands systems for control of agricultural wastewater can be designed on the basis of information from the many municipal wastewater treatment systems in operation all over the world and a few agricultural systems in North America and Australia.

System Components

Emergent Marsh: A shallow basin with densely growing marsh vegetation–typically cattail (*Typha*), bulrush (*Scirpus validus or cyperinus*) reed (*Phragmites*), or rushes (Juncus, @ieoc@aris)-in 1 0-20 cm of water. The marsh functions to remove organic load (BOD_5, suspended solids (TSS), and pathogens as well as in ammonification.

Pond: A constructed pond with 0.5-1 m water depths similar to an aerobic lagoon. Duckweed (*Lemna*) grows on the surface of the pond and various algae within the water column.
Submerged pondweeds with linear, filiform leaves (*Potamogeton, Ceratophyllum, Elodea, Vallisneria*) may be planted in shallow portions of the pond to increase microbial attachment surface area. The pond functions in further reduction of BOD_5 and most significantly for nitrification and denitrification.

Meadow: The meadow is constructed and may be operated similar to an overland flow system. It is planted with reed canary grass (*Phalaris arundinacea*) or other water tolerant grasses and sedges, and it receives the effluent from the pond as shallow sheet flow distributed across the width of the meadow cell. When operated under continuous flows, water depths are maintained at 1-5 cm, but if operated as an overland flow system, wastewater is batch applied to one cell flowing across and down the cell to drain off the lower end while the alternate cell is allowed to dry and oxidize. In either case, the meadow functions for removal of TSS (primarily algae) generated in the pond and for nitrification and denitrification.

Comparatively, the marsh functions most efficiently for BOD_5, TSS, and pathogen removal, but the pond and overland flow meadow, because of the greater amount of oxidized environment, are more efficient at transforming ammonia to nitrogen gases. However, a lightly loaded marsh and meadow will provide similar removal efficiencies, though the total required treatment area may be similar to the combined area in a marsh/pond/meadow system. Primary treatment in a lagoon and a marsh loaded at 100 Kg BOD_2/ha/day will provide treatment to secondary discharge standards - <30 mg/L BOD_5 and TSS and <200 colonies/100 ml fecal coliforms - but limited nutrient control.

DESIGN CRITERIA

Primary treatment should be provided by a single- or multistage lagoon system designed to achieve a 50 percent reduction of the BOD_5 and TSS loading in the wastewater stream. If a lagoon is not present or impractical, a settling basin designed to remove solids, grit, and debris must be located upstream of the wetlands.

Wetlands site selection and delineation will be controlled by the requirement to provide gravity flow for wastewater to the system, between system components, and within each component of the system to eliminate costs and maintenance of pumping wastewaters. Similarly, water control structures or devices should consist of simple "T" pipe along the length of inlet distribution piping and swiveling "elbow" piping or flashboard/stoplog constructs for discharge control structures. Neither clogs as easily or requires adjustment or other complex maintenance typical of ball- or gate-valve flow control devices (Hammer, 1991).

Waste Loading Computations

Determination of effective treatment area required to achieve desired discharge standards is based upon: 1) the quantity (mass) of organic wastes to be treated per day; and 2) the capacity for a given area of wetlands to transform a fixed quantity (mass) per day. Consequently, calculations on the required treatment area begin with determining the total organic load generated at the feedlot, barn, or other facility or in some cases the entire production unit. Representative values of BOD_5, nitrogen, and phosphate production per day from cattle, swine, and poultry are shown in Table 1.

Estimation of waste generation by specific parameter for dairy cattle, swine, and poultry is calculated by multiplying the number of stock by the value (in grams) for the parameter of interest and converting to kilograms (Kg). Since these are average values that do not include waste hay or feed and other sources of organic waste, a prudent designer will include a factor of 10-20 percent to estimate the total organic load generated per day.

A less exact method uses typical concentrations of 2000-4000 mg/L BOD_5 300-500 mg/L NH^3, and 75-150 mg/L total P in raw livestock wastewater and measures the volume of daily flows to estimate the total daily waste loading for each parameter. However, concentrations vary substantially with husbandry practices, type and age of stock, and with seasons of the year.

The total organic load generated per day can then be used to examine the treatment capacity available in an existing anaerobic or aerobic lagoon or to design a new lagoon to provide primary treatment. In either case the lagoon should accomplish a minimum of 50 percent and preferably 60 percent reduction of BOD_5 and suspended solids to reduce the amount of treatment area required in the wetlands system.

Incorporating a lagoon or settling basin for primary treatment provides storage capacity for seasonal application to the wetlands, reduces the treatment area needed in the wetlands, and accomplishes pollutant reduction more efficiently than a stand-alone wetlands system. To illustrate this, a plot depicting pollutant removal efficiency or concentration levels on the Y axis and retention time (an indirect volumetric value) or capacity as volume or effective treatment area on the X axis produces an exponentially decreasing curve with the line for lagoons nearest the Y axis, the wetlands line intermediate, and the overland flow line farthest from the Y axis. Almost irrespective of the treatment method used, the greatest reductions in organic loading and solids occurs at the highest initial concentrations with substantially lower percentage reductions occurring as pollutant concentrations decline. For example, reduction of BOD_5 from 3000 mg/L to 300 mg/L often requires relatively little treatment area or retention time as compared to the treatment needed to reduce 300 mg/L to 30 mg/L. The reduction from 30 mg/L to 10 mg/L requires even more treatment area or retention time. Not only are lagoons more efficient (unit area basis) at high pollutant concentrations, but levels above 300-500 mg/L would stress a wetlands or overland flow systems, perhaps even to the point of failure. Conversely, wetlands and overland flow systems are much more efficient (unit area basis) at reducing 300 mg/L to 30 mg/L and even more so in reducing 30 mg/L to 10 mg/L. Since most natural treatment methods, including wetlands, produce small amounts of BOD_5 and some solids, levels below 5-10 mg/L are unlikely to be achieved.

Lagoons are also effective at reducing high levels of phosphate to moderate levels. However, lagoons are ineffective at removing ammonia through nitrification and denitrification because ammonification during organic matter decomposition creates ammonia. Consequently, anaerobic lagoon discharges may have ammonia nitrogen levels of 400-500 mg/L, though an in-series aerobic lagoon following an anaerobic lagoon will have lower nitrogen levels. Since the age of an existing lagoon, past and current management practices, and loading influence removal performances, it is imperative that the designer have available information on actual concentrations of contaminant parameters in the lagoon discharge before estimating expected loading on the wetlands system.

Required Treatment Area

The treatment area needed to reduce organic and nutrient levels in lagoon discharge to secondary discharge standards has been determined from wetlands use to treat municipal wastewaters and from transformation/assimilation studies of natural wetlands systems. To achieve a discharge level of 30 mg/L BOD_5 the wastewater loading must not exceed 100 Kg/ha/day (90 pounds/acre/day) of effective treatment area. Similarly for NH_3 discharge levels below 10 mg/L, the total nitrogen (TKN) loading rate must not exceed 10 Kg/ha/day (9 pounds/acre/day). Not surprisingly, phosphorus recommendations are one order of magnitude lower (e.g., 1-1.5 Kg/ha/day) since wastewater concentrations of these substances differ by a reverse order of magnitude. Generally, livestock as do municipal wastewaters have 10 times as much nitrogen as phosphorus and 10 times as much BOD_5 as nitrogen so that the required treatment area for each of these parameters tends to coincide for any specific wastewater stream. Note also that organic loading of 10 times the nitrogen level will insure that minimally adequate carbon is available for desired nitrogen transformations.

The effective treatment area required is calculated by dividing the organic load by 100 to derive the answer in hectares. For example, a wastewater flow with 300 Kg/day would require 3 hectares of effective treatment area in the marshes to produce discharge flow concentrations of <30 mg/L and, similarly, with the other parameters. However, this represents effective treatment area, not necessarily everything within the dikes. The effective treatment area is determined by measuring across the cell from the base of the internal side of the dike, not from the top of the dike, since in large systems the dike network occupies a significant amount of surface area. During periods of low water level operation (5-8 cm), adequate treatment area is still available if this design specification method is used.

A minimum of two marsh cells should be included to allow for removing one from service for maintenance, if necessary. A single pond will be adequate, and a single meadow may be used if the meadow is to be operated under continuous flow. If the meadow is to be operated as an overland flow system, then the system must have two meadow cells to permit alternating application and drying.

Proportionately, after computation of required area for each component—marsh, pond and meadow—the effective treatment area tends to be approximately 50 percent in the marshes, 30 percent in the pond, and 20 percent in the meadow(s).

Computations of Required Treatment Areas-Tertiary Treatment

For example, the following computations are for a dairy herd of 150 cows. Waste load generated per day:

BOD_5 produced per cow per day = 773 grams X 150 cows = 115950 grams = 115.95 kolograms
per day, or 116 Kg/day.

Factoring in waste hay or other feed, etc. = 116 X 1.1 = 127.6 or 128 Kg/day.

Nitrogen produced per cow per day = 186 g X 150 = 27900 g = 27.9 Kg or 28 Kg/day.

Anaerobic lagoon: The lagoon may be loaded at 200 Kg BOD_5/ha/day in regions with limited periods of air temperatures below 0°C or 100 Kg BOD_5/ha/day in colder regions or seasons. Removal efficiency = 50% of BOD_5, 20% of nitrogen.
Effluent -- BOD_5 = 128 X 0.5 = 64 Kg/day; Nitrogen = 28 X 0.8 = 22.4 Kg/day.

Marshes: Waste load = 64 Kg/day.
Application rate = 100 Kg BOD_5/ha/day.
Treatment area = 64 Kg/day / 100 Kg BOD_5/ha/day = 0.64 hectares in two marsh cells.
Marsh N removal = 30% reduction; effluent from marshes = 22.4 X 0.7 = 15.78 or 16 Kg/day in effluent.

Pond: Application rate = 40 Kg N/ha/day.
Required area = 16 Kg N/day / 40 Kg/ha/day = 0.4 ha.
Pond N removal = 60% reduction = 6.4 Kg/day in effluent.

Meadow -- continuous flow: Application rate = 20 Kg N/ha/day.
Required area = 6.4 Kg N/day / 20 Kg/ha/day = 0.32 ha.
Meadow N removal = 90% reduction = 0.6 Kg/day.
Meadows operated as overland flow systems may be loaded at 30-40 Kg/ha/day.

Total Wetlands System Area

Marshes = 0.64 ha; pond = 0.4 ha; meadow = 0.32 ha; Total = 1.36 hectares or 3.4 acres.

Required Treatment Area-Secondary Treatment

Reduction of organic loading, suspended solids, and fecal coliforms to secondary levels may be obtained without the pond and meadow, but discharge levels of nitrogen will be high unless a lower loading rate, e.g., 50 Kg BOD_5/ha/day, is used for the marshes.

Marshes: Waste load = 64 Kg/day.
Application rate = 75 Kg BOD_5/ha/day.
Treatment area = 64 Kg/day / 75 Kg BOD_5/ha/day. 0.85 hectares or 2.1 acres in two marsh cells.
Marsh N removal = 60% reduction; effluent from marshes = 22.4 X 0.4 = 8.96 or 9 Kg/day in effluent

Configuration

Though we tend to design rectangular cells due to ease of drawing and calculating, the wetlands cells could as readily be trapezoidal or semi-circular. However, if irregular designs are used, the widest portion should be located at the inlet end to facilitate equal flow distribution. Regardless of the shape, marsh cells should have a 3-5:1 length to width ratio to reduce excessive loading at the inlet end—as occurs in long, narrow cells—and to provide adequate length to initiate nutrification after most of the BOD_5 and TSS load has been removed. Wide, short cells perform fairly well for BOD_5 and TSS removal but poorly for ammonia removal.

Ponds may be rectangular, square, round, or broadly elliptical, but either extreme in length to width ratios must be avoided unless complex inlet and collector distribution systems are used to prevent short circuiting.

Meadows should be rectangular, with a 3-5:1 length to width ratio or greater if operated as overland flow systems.

Dikes and Water Control Structures

Each component of the wetlands system is basically a shallow pond or lagoon. Design and construction techniques used for farm ponds or treatment lagoons are appropriate for general features such as dikes, berms, and typical flashboard/stoplog water control devices. However, dike freeboard must accommodate an organic matter (peat) accumulation rate of 2-3 cm/yr in the marshes and should extend 75 cm above normal water level elevation for each 10 years of projected operation. Adequate freeboard and water level control is also necessary to provide capacity for flow beneath expected thickness of ice cover in colder climates. Piping from the lagoon must be directed to a splitter box incorporating simple weir or flashboard devices to adjust the inlet flow into each marsh cell. The inlet pipe for each marsh cell and the meadows should intercept the inlet distribution pipe at its midpoint in a "T" configuration, and the distributor pipe must extend across 90 percent of the width of the cell. Inlet distributor pipes must be level and supported by concrete stands 30-45 cm above the substrate. "T" pipe fittings equidistant along the length of the inlet distribution pipe facilitate precise adjustment of flows from each "T" ensuring even distribution of wastewaters across the width of the cell. Large gravel or rock (10-15 cm) should be placed immediately below and in front of the inlet distributor to reduce erosion.

Collector piping at the effluent end of each marsh cell and the meadow should be slightly below the cell bottom in a gravel lined trench extending 90 percent of the width of the cell. The discharge pipe may intercept the collector pipe at any convenient location but must terminate in a weir/flashboard structure or an "elbow pipe" mounted with an "O" ring to permit swiveling of the elbow up or down to maintain desired water depths in the marsh or meadow cells.

Bottom Form and Liners

Bottom slopes for the marshes, pond, and meadow should be essentially flat. Width slope for the marshes and meadow must be flat to ensure equal flow distribution of wastewaters. Length slope may not be $>0.05\%$ in the marshes and continuous flow application meadows. Meadows designed for overland flow operation should have slopes of 2-3 percent.

Each component of the system should have an impermeable (hydraulic conductivity of <10 cm/sec) lining, whether formed by compacting in situ materials or introducing bentonite or geotextile fabrics, since nitrate/nitrite levels may be high enough to cause concern for potential groundwater contamination as far down as the midpoint of the meadows.

Construction

Careful supervision of contractors is imperative to insure that grade and elevation specifications are precisely met. Otherwise, considerable difficulty with short-circuiting and reduced treatment capacity may occur and be impractical to correct during operation.

Vegetation

Marshes should be planted with cattail or bulrush on 1 meter centers during the first half of the growing season. Pondweeds in weighted cotton mesh bags and duckweed may be simply placed in the pond. Perennial grasses suitable for the meadow include Reed Canary, Tall Fescue, Redtop, Kentucky Bluegrass, Orchard Grass, Common Bermuda, Coastal Bermuda, Dallas Grass, and Bah ia. Marsh planting materials may be dug locally if suitable digging methods are used. Within a natural stand of cattail or bulrush, remove the stems, rhizomes, and roots from a 0.5 m^2 area, and then move over 1 meter before renewing digging. If this method is used before the midpoint of the growing season, depleted areas will be recolonized by the end of the growing season with little impact on natural wetlands.

Planting materials may also be obtained from commercial sources, but in either case, each root stock must include 20-25 cm of stalk after removal of the tops. Failing to remove tops will result in wind-throw before the roots have become established, and lack of a short stalk protruding above the surface of the water will cause plant mortality if wastewater with little dissolved oxygen is applied to the cell (Hammer, 1991).

Depending upon labor costs, digging local materials may be more costly than purchasing plant stocks from a nursery or supplier. Generally, planting materials may be purchased for $150/thousand from Wildlife Materials in Oshgosh, Wisconsin, and planting costs are $5,000-6,000 per hectare. Digging locally and planting has cost up to $12,000 per hectare. Sago or other pondweeds (linear, filiform leaves) and tapegrass are usually planted as tubers and simply weighted in mesh bags and dropped into the water or placed in soft, wet muds at desired locations.

Local digging is facilitated by using a back hoe that removes the top 15-20 cm of substrate and plant materials. Similarly, a trencher may be used to dig shallow trenches perpendicular to the long axis of each cell with plants set into the trench and soil damped around them.

Planting bulrush or cattail is similar to planting any garden plant, but careful supervision is important because plants with damaged roots or plants that are placed incorrectly will not survive. Shallow flooding followed by draining, which leaves "soupy" substrates, creates ideal growth conditions for new plant stocks and facilitates hand or mechanical planting. The area should be flooded to 1-5 cm after planting, but water levels must not overtop cut stalks from the original plants or new shoot growth. After all planting is finished, water levels should be gradually raised to normal operating elevations as the plantings grow higher, but water levels must not overtop new growth during the first growing season. Emergent wetlands plants are not as susceptible to drowning after the first growing season or in waters with relatively high dissolved oxygen content.

Alternatively, cattail, and to some extent the other species, can be established by simply manipulating the water levels at the appropriate time of the year. Or cattail and others may be seeded, but germination rates are very low, e.g., 3-5 percent. Unfortunately, this method is dependent on natural means of seed dispersal and germination and may require more than one growing season to develop a dense stand.

Grasses may be sown following standard recommendations. Full strength wastewater should not be applied until grass height has exceeded 15-25 cm depending upon the type and variety.

Mosquito fish (Gambusia) and top minnows (Pimephales) or other bait fish may be introduced into the marsh and pond after operating water depths have been stabilized. Bottom-feeding fish (e.g., carp, catfish) should not be used, since low turbidity water is important to various treatment processes.

Operation

Equal flow distribution across the width of the marsh and meadow cells is obtained by adjusting the angle of each "T" outlet in the inlet distributor piping. Adjustment is done by inserting a lever (short board) into each "T" and gently rotating the "T" to the proper elevation. Discharge control structures (flashboard/"elbow pipe") must be adjusted to maintain 10-20 cm water depths in the marsh cells, 0.8-1.3 m in the pond, and 1-5 cm in the meadow.

Routine weekly inspections are necessary to ensure equal flow from each "T" outlet on the inlet distributor piping. If clogging occurs, rotate the "T" downward to drain the pipe, remove the obstruction, and rotate it upward to the desired operating angle. Plant debris obstructing outlet control structures may also need to be removed by similarly rotating the elbow piping or raking from a flashboard device. Water levels in each component should be checked and adjusted as necessary and all piping visually inspected for cracks or leaks. Dikes and flow control structures should be inspected for leaks and corrective action implemented.

Flow distribution within cells should be occasionally inspected to detect channel formation and short-circuiting and corrected by planting vegetation or filling soil in any channels. Vegetation should be visually inspected for signs of disease (yellowing/browning, spots, wilting, etc..), insect infestations, or stress (stunted growth) during other inspection periods.

Shrubs or trees must be removed from the wetlands cells because they will shade out desirable plant species. Weeding of herbaceous plants is unnecessary as is harvesting. Accumulated leaf and stalk litter creates a desirable layer of humic materials on the surface of the cells within which much of the wastewater treatment occurs. Foot traffic should be minimized and vehicular traffic prevented within the cells because either will compress and damage the humic/compost surface layer. Pesticides or other chemicals that may harm the vegetation should not be directly applied or introduced into the wastewater stream.

NUTRIENT/SEDIMENT CONTROL SYSTEM

The nutrient/sediment control system combines marsh/pond components of constructed wetlands with other erosion/sediment management elements to use physical, chemical, and biological processes for removal of sediment and nutrients from runoff. This system was originally designed for potato fields in northern Maine, but it could be easily adapted to pasture or crop field runoff as well as to urban stormwater runoff in other regions. Though it may be used in any small watershed, system performance and longevity will be considerably enhanced if standard erosion control practices - grassed waterways, terraces, no-till cultivation, sediment basins, filter strips, diversions, etc. - are put in place prior to locating the nutrient/sediment system. Land area requirements range from 0.5-1.4 ha for 23-68 ha watersheds, respectively. In addition, functioning is not dependent upon a minimum critical size, so a number of units can be judiciously located in small or large watersheds to accomplish treatment of virtually all runoff.

Components of the system include a sediment basin, level-lip spreader, primary grass filter, wetland (marsh), deep pond, and a polishing filter (Figure 7). The sediment basin is a trapezoidal trench designed to collect larger sediment and organic particles and to regulate flow to the remainder of the system. The level-lip spreader consists of a trench filled with crushed rock or sorted stone that provides sheet flow to the third component. The primary grassed filter is a modified overland flow unit with lower gradient, broader width, and tile subsurface drains for fine sediment and nutrient removal. The freshwater wetland consists of an emergent marsh similar to the constructed wetlands described above, which along with the deep pond (2-5 m) functions primarily for nutrient removal. The final component, a polishing filter, generally consists of a natural wet meadow, shrub, or wooded area for removal of algae and some nutrients. If excessive runoff is anticipated (i.e., an urban watershed), a small retention basin above a smaller sediment ditch or in leu of the ditch may be added, but the basin/ditch and level-lip spreader must provide sheet flow to the primary grass filter for proper operation.

These systems are typically situated at the downstream end of grassed waterways or other small water courses prior to their junction with intermittent or permanent streams, rivers, and lakes, Consequently, an individual farm may require a number of small, judiciously placed units to accomplish complete control of nutrient and sediment removal. However, the minimal acreage and flexible design requirements are amenable to a variety of topographic, land use, soil, and meteorological conditions. Sizing is related to size of watershed and anticipated runoff (Table 2).

Plant species include grasses appropriate to the region that will provide good ground cover during the period of highest expected runoff (i.e., cool season grasses in northern regions and fescue or Bermuda strains in the southeast). Cattail, bulrush, or rushes may be used in the marsh and the transition zone with the pond supports submergent wetlands species-pondweeds, tape grass, etc.

Recommended maintenance consists of removal of accumulated sediment in the ditch and grass from the primary grass filter. If extensive algal mats develop in the pond, their removal during maximum growth will enhance nutrient removal. In addition, if the pond is used for bait fish production, it will enhance nutrient transport out of the system and financial return to the landowner. Obviously, larger fish, especially bottom feeders such as carp or catfish, would be detrimental to sediment removal and biological processes and should not be used.

In practice, nutrient/sediment control systems have removed 90-100 percent of suspended solids, 85-100 percent of total phosphorus, 90-100 percent of BOD_5, and 80-90 percent of total nitrogen from potato field runoff in northern Maine. Construction costs have ranged from $13,500 to $23,000 for 8 ha and 67 ha watersheds, respectively (Wengryznek, pers. comm.).

These costs, amortized over the expected life of the system, plus annual maintenance average only $50-$60 per ha per year for the contributing watershed. This is a very favorable factor when compared to the potential costs of less effective and non-productive conservation practice alternatives.

Nutrient/sediment control systems should not be used to replace conventional BMPs but should complement a sound land and water management program in each watershed.

OTHER CONSIDERATIONS IN CONSTRUCTED WETLANDS TREATMENT SYSTEMS

Advantages

Advantages of constructed wetlands include relatively low construction costs—essentially grading, dike construction, and vegetation planting with little steel or concrete—and low operating costs. Maintenance consists of monitoring water levels and plant vitality, collecting NPDES samples, and grounds maintenance (mowing dikes and roadways). Properly designed and constructed systems do not require chemical additions, internal pumping, sludge handling, or other procedures of conventional treatment systems. Neither do they require plant harvesting, except in specialized applications using floating plants—water hyacinth (Eichhornia) or duckweed (Lemna) for nutrient removal after conventional treatment. In these cases, maintenance costs may be very high.

Typically construction costs range from 1/10 to 1/2 of the cost of comparable conventional treatment systems. For example, a TVA designed system at Benton, Kentucky, which polishes primary lagoon effluent, cost $260,000 in 1986 compared to a 1972 estimate of $2.5 million for a comparable conventional treatment system. Two other systems designed for secondary and tertiary treatment for communities of 500 (Hardin) and 1,000 (Pembroke) users varied from $212,000 to $366,000. Operating costs for these systems are less than $10,000 per year. A TVA wetlands that controls acid mine drainage cost $28,000 to construct and plant—about the same as the costs for chemicals alone to provide comparable treatment for only 1 year (Brodie, et al, 1988). Operating costs, other than monitoring sample collection and analysis, have been less than $500 per year.

Wastewater treatment efficiencies are very good, especially for BOD_5, TSS, and fecal coliform bacteria with common discharge values of 10-20 mg/L and 50-100 colonies per 100 ml. With proper design and adequate treatment area, removal of nitrogen compounds and phosphorus are readily accomplished. As can be expected, performance varies with different designs, wastewater sources, amount, and type of pre-treatment- and treatment-area/retention times, with the most variation related to the type of system and treatment-area/retention times. Constructed wetlands are also amenable to substantial fluctuations in loading rates, adapting to weekly and annual fluctuations in flows. This adaptation of constructed wetlands to fluctuations in loading can be observed, for example, from a high school in northwest Alabama.

Constructed wetlands can provide ancillary benefits in the form of wildlife habitat, recreational and environmental space, or simply urban greenspace. Recreational activities derive from vertebrates, larger invertebrates, and, to some extent, vegetational components. For example, the Arcata system in California has been described as a bird watching "hotspot" in a national "birding" publication (Gearhart et al., 1989). During a Sunday visit to the Martinez, California, wetlands system, although available police and janitorial employees were unable to direct me to the wastewater treatment plant, a chance encounter with a local birdwatcher saved the day. Amateur naturalists and environmental educators are able to quickly identify and exploit the educational and recreational benefits of a simulated marsh nearby (James and Bogaert, 1989). Systems located near urban areas may also provide greenspace benefits (Smardon, 1989) or simply open, natural areas that attract a variety of low-intensity recreational users - walking, jogging, picnicking, relaxing, etc. Treatment system operators, pleased with the attention and support received from local citizens, usually welcome ancillary uses. More importantly, many realize that recreational benefits and pleasing aesthetics of wetlands systems may reduce opposition to new or expanded systems.

However, wetlands constructed for wastewater treatment, at least initially, are comparatively simple, often monotypic species systems. A properly designed and constructed cell with adequate treatment area that is covered in a dense stand of Typha or Scirpus will efficiently remove target contaminants from influent waters while providing habitat for a few muskrats, blackbirds, and some songbirds but little else (Hammer, 1989b). Even if operated at maximum efficiency, however, it will not have adequate capacity to store flood waters or release substantial quantities to amplify low stream flows in dry conditions. Wastewater treatment has been maximized through optimized design and operating criteria, and all other functional values have been subordinated. But the water improvement function is still efficient and enduring even though other wetlands functional values are substantially reduced or nonexistent.

Disadvantages

Constructed wetlands are land intensive—they require much more land area than do package treatment plants—and they require relatively level surfaces. However, nutrient/sediment systems need relatively less land area and may be constructed in watershed locales with considerable relief. Where land costs are high such as in larger cities or in very rugged terrain that would need considerable cut and fill constructed wetlands are more expensive to construct than conventional systems, although lower operating costs over a 20 or 30 year plant lifetime must be factored into the decision process. Current mass-loading design recommendations require 6-20 ha of treatment area per 4,000 m^3, depending upon the level of pre-treatment and the desired discharge limits.

In addition, present design, construction, and operating criteria are imprecise—the reason for the range of treatment area requirements in the above. And wetlands systems, either natural or constructed, are complex, dynamic systems about which we have only limited understanding. However, a number of experimental and operating systems throughout the world are beginning to accumulate the data base from which precise design and operating criteria will be developed.

Another disadvantage is delayed operational status. Because peak removal efficiencies of constructed wetlands are dependent upon vegetation growth and establishment, design efficiencies are not likely to be attained until after two or perhaps three growing seasons. Completing construction and planting does not immediately translate into full operational status, and potential users must plan to gradually phase in wetlands operation concurrent with phase out of the existing conventional system. Proper operation of the existing plant during this phase can be critical to avoid discharging activated sludge during high flow periods or simply "wasting" sludge to the wetlands system before it is fully operational.

Treatment system longevity is poorly documented, since no successful operating-scale system has been in operation for more than 20 years. Because these systems simulate natural wetlands ecosystems that have functioned to purify water for thousands of years, I expect that system efficiency is not likely to be detrimentally impacted by age, but artificial constraints may require modifications or restarting after some period of time. Litter/detritus accumulation rates have been measured at about 2 cm/year in municipal systems with no loss of treatment efficiencies (Haberl and Perfler, 1989). Therefore, designs should incorporate this accumulation factor in dike height specifications and dikes should have 50-55 cm of freeboard for a 20-year operating lifetime, or greater for longer operational status. At that point, the system may need to be cleaned out and restarted, and after testing to identify possible toxic substances, accumulated litter may be composted or land applied similar to conventional sludge,

Finally, improperly designed or operated constructed wetlands could create pest problems mosquitoes or rodents-that may cause adverse impacts to local residents. Both are easily avoided with appropriate designs and operating procedures (Martin and Eldridge, 1989).

Toxics Accumulation

Since wetlands treatment systems transform or remove pollutants from inflowing waters, the ultimate fate of certain substances within the wetlands ecosystem is of more than academic interest (Hammer, 1990). Depending upon the source, influents may contain various natural and anthropogenic organic compounds, metals including heavy metals, pathogenic organisms, salts, etc. A few materials (e.g., selenium) are selectively taken up by plants, but most are precipitated or complexed within the substrate. Generally, only 4-5 percent of the nutrient loading on a wetlands system is incorporated into plant or animal tissue. However, some metals may occur in relatively high concentrations. For example, iron levels as high as 5000 mg/kg and manganese levels up to 4100 mg/kg were present in cattail leaves and stems grown in experimental cells that were heavily loaded with acid mine drainage. Only traces of other metals were present (TVA, unpublished data). Copper was nonexistent in cattail from a natural marsh but averaged 6.1 mg/kg in cattail from two municipal wastewater treatment systems. Higher concentrations of lead were found in a natural cattail stand (1.7 mg/kg) than in cattail from the municipal systems (0.3 gm/kg).

Though only low levels of potential toxic metals occurred in these samples, long-term effects of relatively high levels of iron and manganese are not known. In the short term, iron and manganese did not appear to have detrimental effects on cattail growth and vitality in the experimental cells. In fact, plants in the upstream portion of each cell were more robust than plants in the lower sections. Upper portions of each cell received raw inflowing acid mine drainage that probably contained small concentrations of micronutrient in addition to substantially higher concentrations of iron and manganese. Differential robustness within each cell was likely due to micronutrient uptake in the upstream portion and limited micronutrient availability to plants in the lower sections.

General Considerations

Because constructed wetlands are open, outdoor systems, they receive inputs of animal and plant life from adjacent areas and from distant sites. Over time are likely to become more and more similar to other naturally occurring wetlands in a region (Hammer and Bastian, 1989). Though we may design and build a system with a specific substrate and only one or two plant species currently thought to be highly efficient, over time, many types of plants and animals will take up residence. Consequently, a constructed wetlands is likely, to become more similar to a natural wetlands as the system matures and ages. To prevent these invasions and attempt to maintain a monoculture would be difficult and costly and may be self-defeating. Living organisms that become established in an operating system are not likely to detrimentally impact treatment efficiency and may very well improve system operation. In addition, maintaining a monoculture is difficult, as any farmer knows, since the single species stand or monoculture is often susceptible to disease, insects, or grazing animals. For example, cattail in a marsh that was treating acid water seepage from coal ash storage ponds was devastated by an outbreak of armyworms during the first year of operation (Snoddy et al., 1989).

Much can be learned from both natural wetlands receiving wastewaters and from constructed systems with a few years of operating history. Despite some attempts to reduce wetlands treatment systems to minimal components and simplify treatment areas by using the most efficient combinations of substrate, vegetation, and loading rates, most successful systems are indistinguishable on casual examination from natural marshes. In fact, poorly performing systems that I have visited did not appear to be viable marsh ecosystems. Generally, the absence of an important component, attribute, or characteristic was obvious to anyone with experience in natural marsh ecosystems. Conversely, successful systems are often quite similar to a natural marsh, and it's beginning to appear that the basis for design of wetlands constructed for wastewater treatment should be to simulate the structure and functions of a natural marsh ecosystem (Hammer, 1991).

This is not to suggest that natural wetlands are casually useable for wastewater treatment. We know too little of the complex interactions among a myriad of components in natural systems and too little of optimal treatment area requirements or application rates to risk damaging natural wetlands. These systems are too valuable to lose, since research on natural wetlands systems will continue to increase our understanding and ability to design and build constructed wetlands for specific purposes. Intact natural wetlands also provide a host of other benefits to society.

Many are searching for an inexpensive, efficient wastewater treatment process, and constructed wetlands are an attractive alternative. But constructed wetlands are not a panacea for all wastewater treatment problems. Although experimental work has been underway for more than 30 years, the technology is still in its infancy and much remains to be learned on design, construction, and operation (Hammer, 1989c). Most previous work has been directed towards municipal wastewater treatment, and our information is adequate to conservatively design and operate systems for that use. Within the last 5 years a number of experimental and operating systems treating acid mine drainage have provided similar information. But the substantial potential for treating NPS pollution especially urban stormwater runoff and agricultural wastewaters, industrial wastes, and even failed septic tank drain fields at individual home sites remains to be developed.

Wetlands accomplish water improvement through a variety of physical, chemical and biological processes operating independently in some circumstances and interacting in others. Vegetation obstructing the flow and reducing the velocity enhances sedimentation, and many substances of concern are associated with the sediment because of clay particle adsorption phenomena. Increased water surface area for gas exchange improves dissolved oxygen content for decomposition of organic compounds and oxidation of metallic ions. But the most important process is similar to decomposition occurring in most conventional treatment plants—only the scale of the treatment area and composition of the microbial populations is likely to be different.

In both cases an optimal environment is created and maintained for micro-organisms that conduct desirable transformations of water pollutants. Wetlands systems use larger treatment areas to establish self-maintaining systems that provide environments for similar microbes but also support additional types of micro-organisms because of the diversity of microenvironments. The latter, along with a larger treatment area, frequently provide more complete reduction and lower discharge concentrations of water-borne contaminants. Regardless, most removal or transformation of organic substances in municipal wastewaters or metallic ions in acid mine drainage is accomplished by microbes—algae, fungi, protozoa, and bacteria.

Conventional wastewater treatment systems require large inputs of energy, complex operating procedures, and subsequent costs to maintain optimal environmental conditions for microbial populations in a small treatment area. The low capital and operating costs, efficiency, and selfmaintaining attributes of wetlands treatment systems result from a complex of plants, water, and microbial populations in a large enough land area to be self-sustaining. It may be less costly to construct a minimally sized, least-component wetlands treatment system, but operational costs to maintain that system could easily negate initial cost savings. However, for small communities, farms, mines, and some industries, a conservatively designed and biologically complex system may provide more efficient treatment, greater longevity, and reduced operating requirements and costs.

REFERENCES

Amaral, M. R., 1989. The Impact of the Floods of the Pantanais Matogrossenses. Comissao de Defesa do Pantanal, Rua Cuiaba No. 440, Corumba, MS, Brazil. 30pp.

Anon, 1991. Nutrient/Sediment Control System (Draft Interim Standard and Specification). CODE I-05. USDA-SCS. Orono, ME.

Brix, H. and H.-H. Schierup, 1989. Danish experience with sewage treatment in constructed wetlands. pp. 565-573. In: D. A. Hammer (ed.), Constructed Wetlands for Wastewater Treatment. Lewis Publishers, Inc., Chelsea, MI.

Brodie, G. A., D. A. Hammer, and D. A. Tomljanovich, 1988. Man-Made Wetlands for Acid Drainage Control in the Tennessee Valley. Proc. Mine Drainage and Surface Mine Reclamation, 1:325-331, Bur. Mines Inf. Cir. 9183.

Cadavid, E. A., 1989. Conservation Plan for the Upper Paraguay Basin, Brazil. Secretarial Estadual de Meio Ambiente, Governo do Estado de Mato Grosso do Sul, Campo Grande, MS, Brazil. 69pp.

Gearheart, R. A., F. Klopp, and G. Allen, 1989. Constructed free surface wetlands to treat and receive wastewater: Pilot Project to full scale. pp. 121-137. In: D. A. Hammer (ed.), Constructed Wetlands for Wastewater Treatment. Lewis Publishers, Inc., Chelsea, MI.

Haberl, R. and R. Perfler, 1989. Root-zone system: Mannersdorf-new results. pp. 606-621. In: D. A. Hammer (ed.), Constructed Wetlands for Wastewater Treatment. Lewis Publishers, Inc., Chelsea, MI.

Hammer, D. A., 1989a. Constructed wetlands for treatment of agricultural waste and urban stormwater. In: S. K. Majumdar et al. (eds.), Wetlands Ecology and Conservation: Emphasis in Pennsylvania. The Pennsylvania Academy of Science.

Hammer, D. A., 1989b. Protecting Water Quality with Wetlands in River Corridors. Proceedings of the International Wetland Simposium on Wetlands and River Corridor Management. July 5- 9, 1989, Charleston, SC, The Association of Wetlands Managers, Inc.

Hammer, D. A., (ed.) 1989c. Constructed Wetlands for Wastewater Treatment. Lewis Publishers, Inc., Chelsea MI. 831pp.

Hammer, D. A. and R. K. Bastian, 1989. Wetlands ecosystems: Natural water purifiers? pp. 5-19. In: D. A. Hammer (ed.), Constructed Wetlands for Wastewater Treatment. Lewis Publishers, Inc., Chelsea, MI.

Hammer, D. A., B. P. Pullin, and J. T. Watson. 1989. Constructed Wetlands for Livestock Waste Treatment. Proceedings of the National Nonpoint Conference, St. Louis, MO, April 23-26, 1989.

Hammer, D. A., 1990. Water Improvement Functions of Natural and Constructed Wetlands. Proceedings of the Newman Teleconference Seminar Series - Protection and Management Issues for South Carolina Wetlands, pp. 129-157. Clemson University, March 28, 1990.

Hammer, D. A., 1991. Creating Freshwater Wetlands. Lewis Publishers, Inc., Chelsea, MI. (in press).

Higgins, M., 1991. The Use of Constructed Wetland Systems in Treating Agricultural Runoff: 1990 Data Summary. Report of the Department of Civil Engineering, Univ. of Maine, Orono, ME.

James, B. B. and R. Bogaert, 1989. Wastewater Treatment/Disposal in a Combined Marsh and Forest System Provides for Wildlife Habitat and Recreational Use. pp. 597-605. In: D. A. Hammer (ed.), Constructed Wetlands for Wastewater Treatment. Lewis Publishers, Inc. Chelsea, MI.

Kadlec, J. A., R. H. Kadlec and C. J. Richardson, 1974. The Effects of Sewage Effluent on Wetland Ecosystems. Research Applied to National Needs. Grant GI-34812X, Univ. of Michigan, Ann Arbor, MI.

Martin, C. V. and B. F. Eldridge, 1989. California's experience with mosquitoes in aquatic wastewater treatment systems. pp. 393-398. In:

D. A. Hammer (ed.), Constructed Wetlands for Wastewater Treatment. Lewis Publishers, Inc., Chelsea, MI.

Miller, G., 1989. Use of Artificial Cattail Marshes to Treat Sewage in Northern Ontario, Canada. pp. 636-642. In: D. A. Hammer (ed.), Constructed Wetlands for Wastewater Treatment. Lewis Publishers, Inc., Chelsea, MI.

Mitsch, W. J. and J. G. Gosselink, 1986. Wetlands. Van Nostrand Reinhold Company, New York, NY. 539pp.

Odum, H. T. and S. Brown, 1976. Regional Implications of Sewage Effluent Application in Cypress Domes. Freshwater Wetlands and Sewage Effluent Disposal Symposium. May 1976. Univ. of Michigan, Ann Arbor.

Pullin, B. P. and D. A. Hammer, 1989. Comparison of Plant Density and Growth Forms Related to Removal Efficiencies in Constructed Wetlands Treating Municipal Wastewaters. 62nd Ann. Conf., Water Pollution Control. Federation, San Francisco, CA, Oct. 1989.

Reddy, K. R. and W. H. Smith (eds.), 1987. Aquatic Plants for Water Treatment and Resource Recovery. Magnolia Publishing, Orlando, FL. 1032pp.

Reed, Sherwood C. (ed.), 1990. Natural Systems for Wastewater Treatment - Manual of Practice FD-16. Water Pollution Control Federation, Alexandria, VA. 270pp.

Seidel, K., 1971. Macrophytes as functional elements in the environment of man. Hydrobiology, 20(1): 137-147.

Small, M. M., 1977. Natural Sewage Recycling Systems. Brookhaven National Laboratory Report 60530, Upton, NY.

Smardon, R. C., 1989. Human perception of utilization of wetlands for waste assimilation, or how do you make a silk purse out of a sow's ear? pp. 287-295. In: D. A. Hammer (ed.), Constructed Wetlands for Wastewater Treatment. Lewis Publishers, Inc., Chelsea, MI.

Snoddy, E. L., G. A. Brodie, D. A. Hammer, and D. A. Tomljanovich, 1989. Control of the armyworm. Simvra henrici (Lepidoptera: Noctuidae), on cattail plantings in acid drainage treatment wetlands at widows Creek Steam-Electric Plant. pp. 808-811. In: D. A. Hammer (ed.), Constructed Wetlands for Wastewater Treatment. Lewis Publishers, Inc., Chelsea, MI.

Table 1. Livestock Waste Production

	Production per Day (grams)			Total Volume (meter^3)
	BOD_5	N	P	
Dairy Cows (455 Kg)	773	186	33	0.05
swine (91 Kg)	180	204	68	0.03
Poultry-layers (1.8 Kg)	6	1.3	0.5	0.000 1
Poultry-broilers (1 Kg)	3.3	0.7	0.3	0.00005

Table 2. Minimum Sizing Criteria for a Nutrient/Sediment Control System for Runoff from Typical Agricultural Rowcrop Fields

Watershed Size* (ha)	Sediment Ditch (m^2)	Grassed Filter (m^2)	Marsh (m^2)	Deep Pond (m^2)	Polishing Filter (m^2)
<10	70	700	925	925	1000
20	90	925	1200	1500	1400
30	115	1200	1400	2000	1700
40	140	1400	1600	2600	2100
60	185	1900	2100	3700	2800

*For larger watersheds, size of the components is increased proportionally to the drainage area.

LIST OF FIGURES

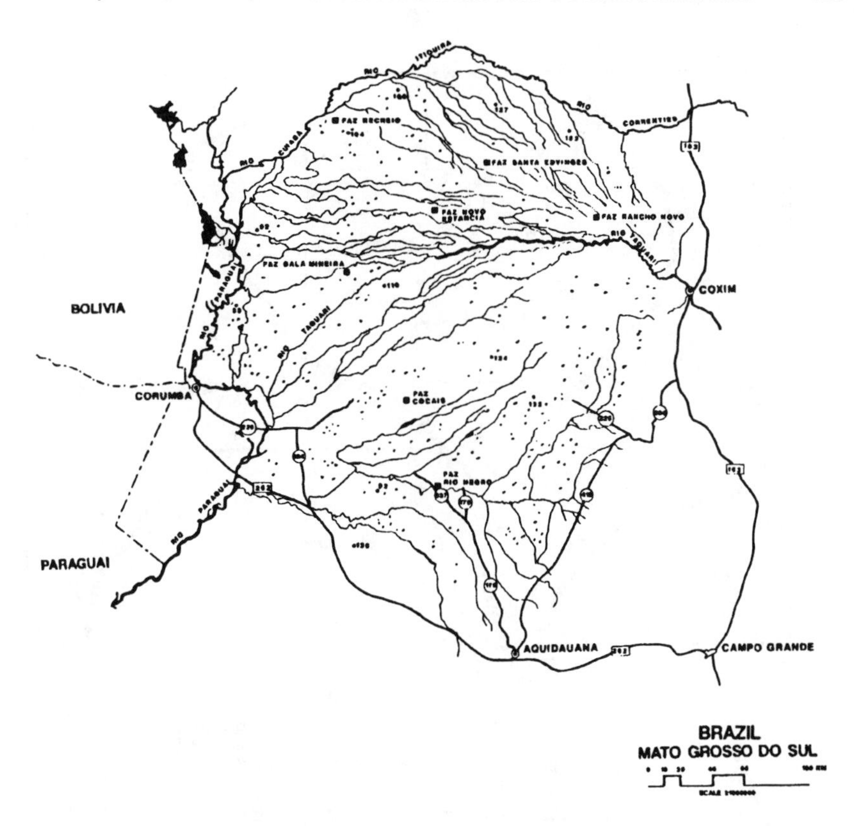
BOLIVIA
PARAGUAI
CORUMBA
COXIM
AQUIDAUANA
CAMPO GRANDE
BRAZIL
MATO GROSSO DO SUL

Figure 1

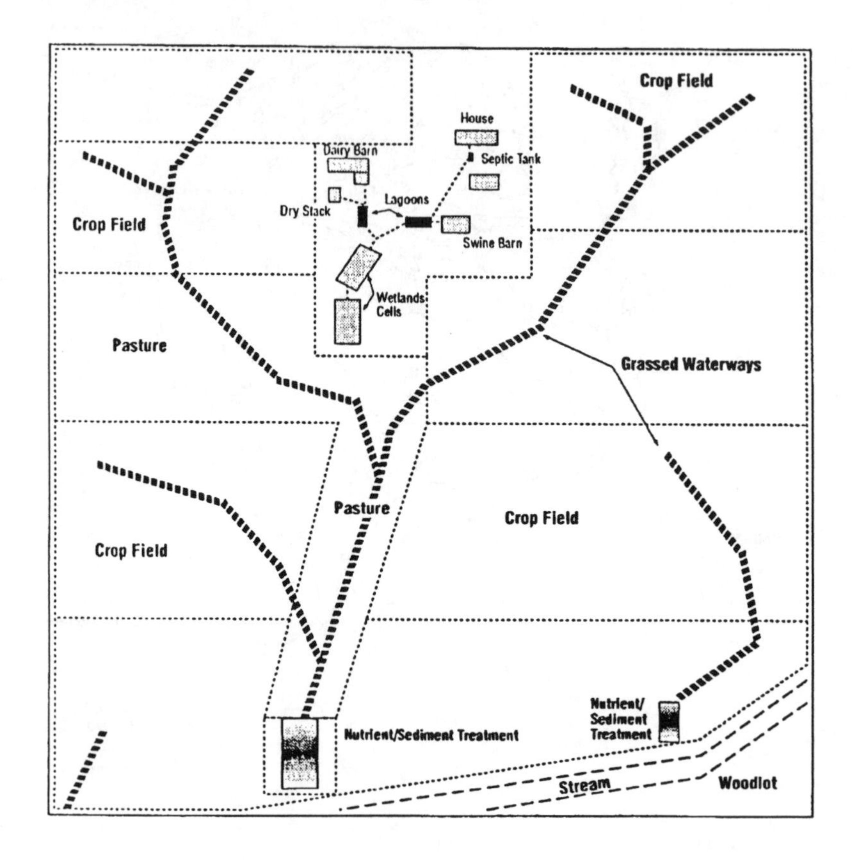
Crop Field
House
Dairy Barn
Septic Tank
Lagoons
Dry Stack
Crop Field
Swine Barn
Wetlands Cells
Pasture
Grassed Waterways
Pasture
Crop Field
Crop Field
Nutrient/Sediment Treatment
Nutrient/Sediment Treatment
Stream
Woodlot

Figure 2

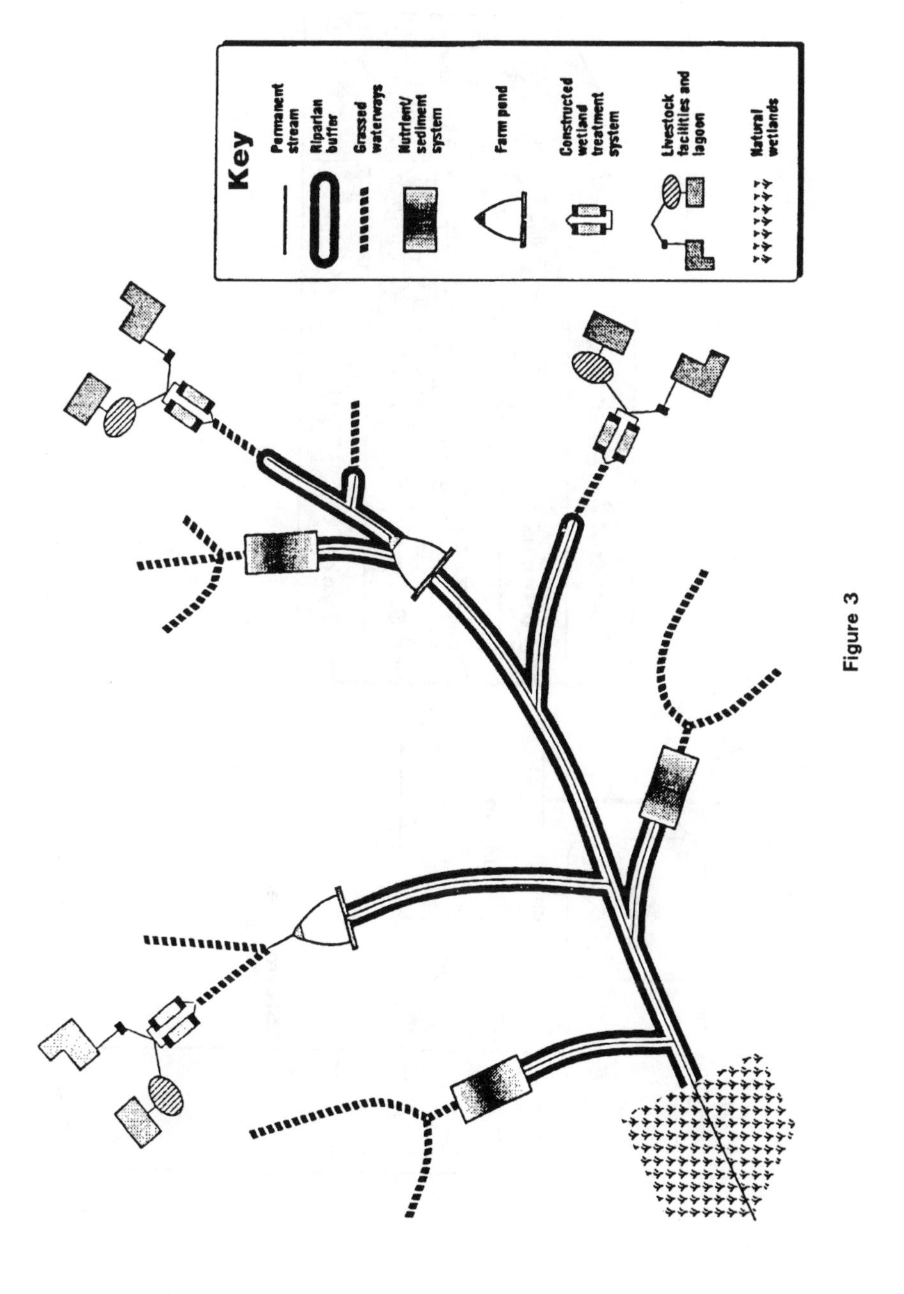
Key
Permanent stream
Riparian buffer
Grassed waterways
Nutrient/ sediment system
Farm pond
Constructed wetland treatment system
Livestock facilities and lagoon
Natural wetlands

Figure 3

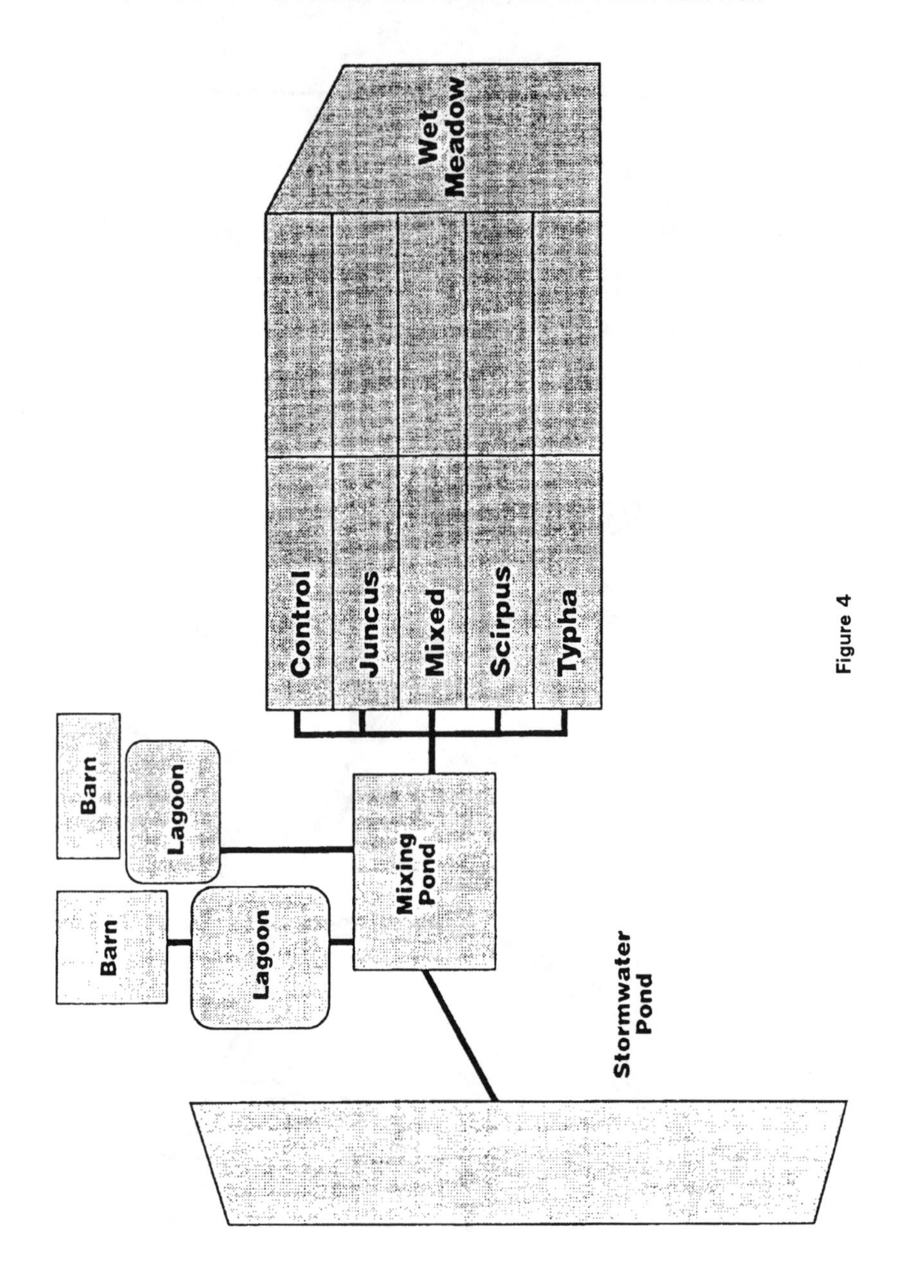
Wet Meadow
Control
Juncus
Mixed
Scirpus
Typha
Barn
Lagoon
Barn
Lagoon
Mixing Pond
Stormwater Pond

Figure 4

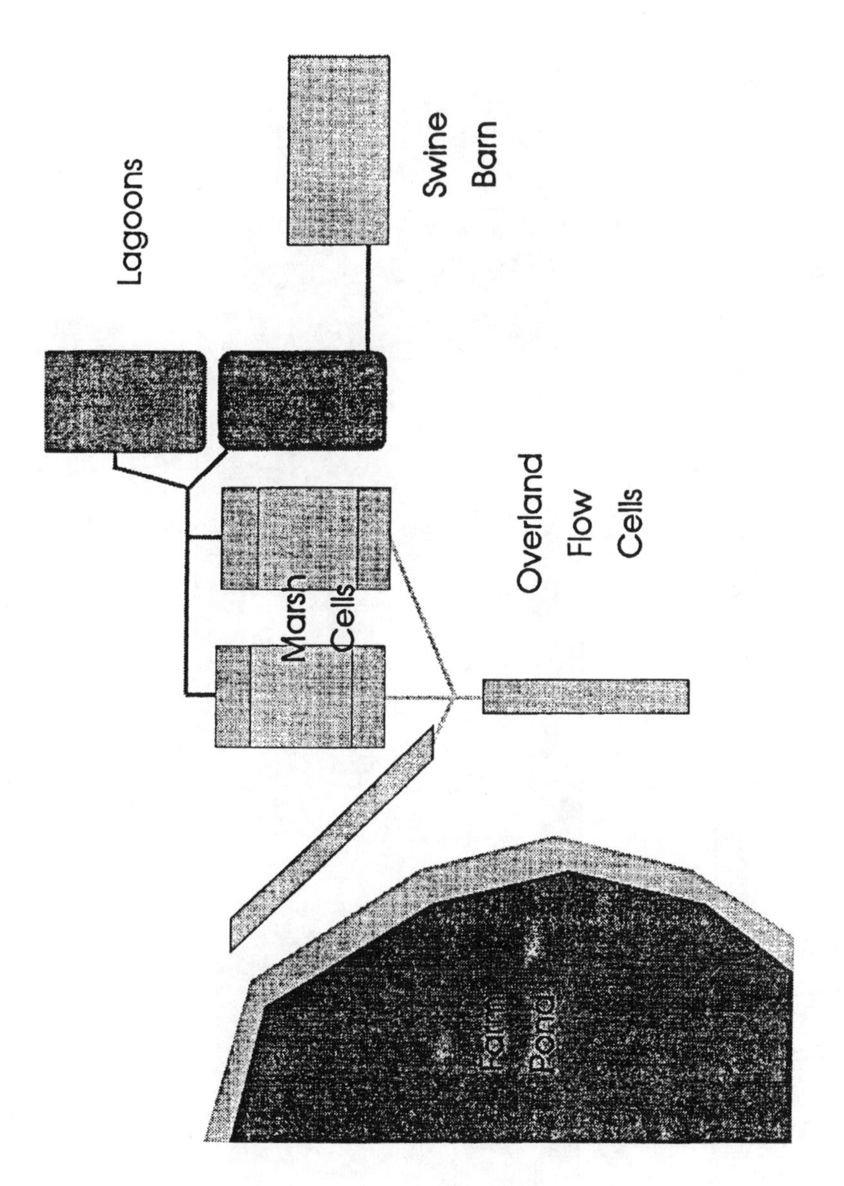
Lagoons
Swine Barn
Marsh Cells
Overland Flow Cells
Farm Pond

Figure 5

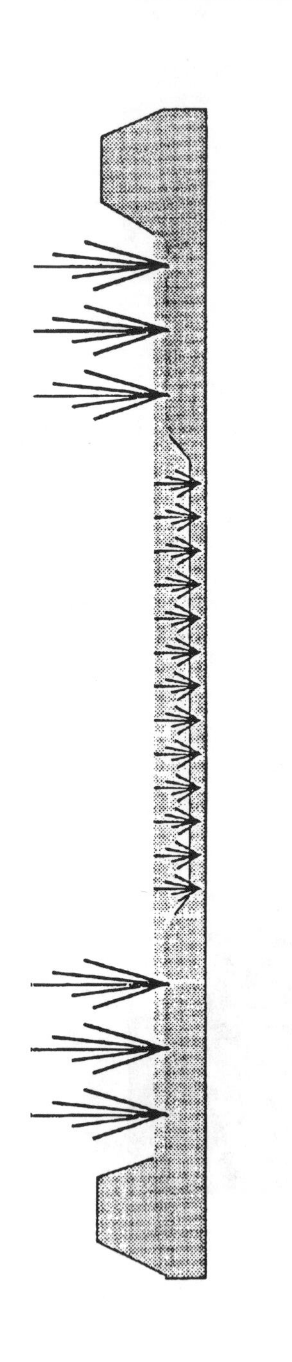

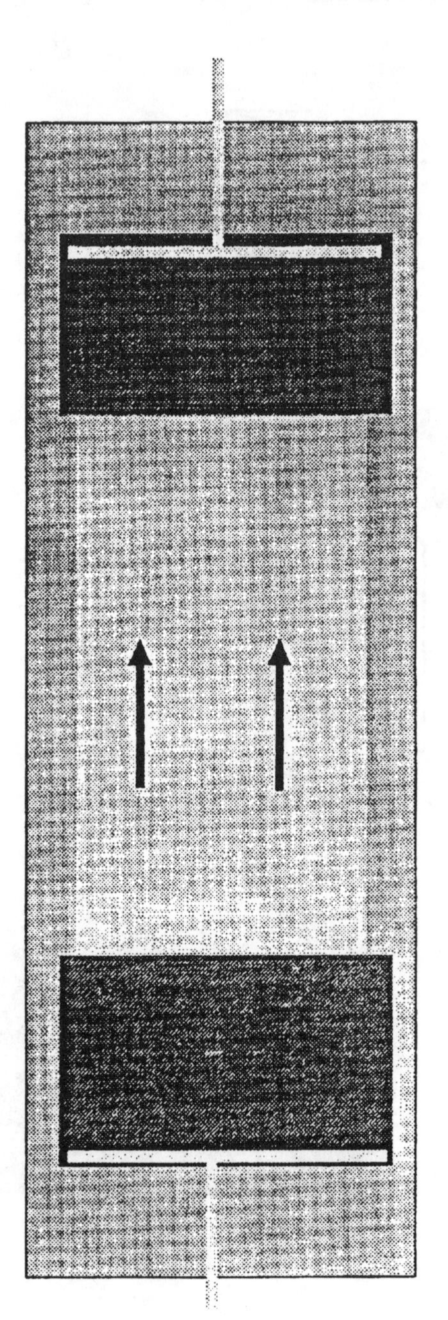

Figure 6

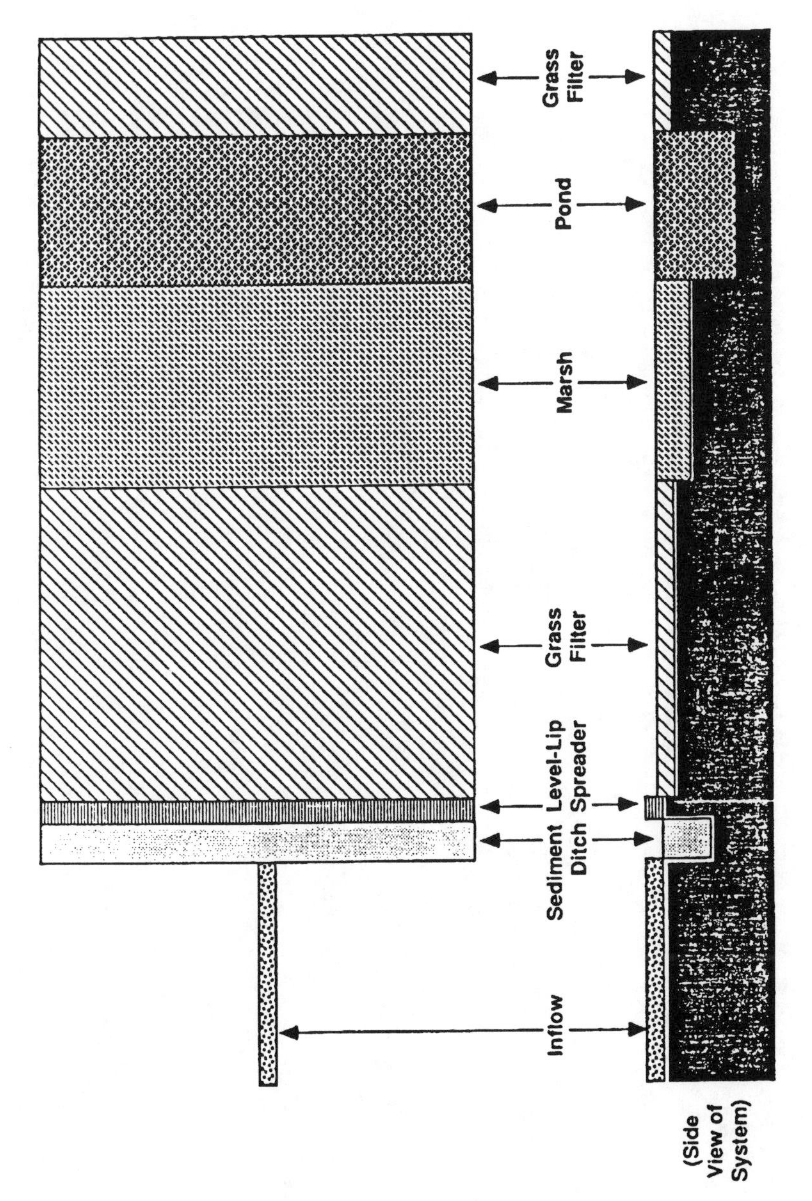
Grass Filter
Pond
Marsh
Grass Filter
Level-Lip Spreader
Sediment Ditch
Inflow
(Side View of System)

Figure 7

CHAPTER 5

Developing Design Guidelines for Constructed Wetlands to Remove Pesticides from Agricultural Runoff

John H. Rodgers, Jr. and **Arthur Dunn**, Department of Biology, Biological Field Station, University of Mississippi.

ABSTRACT

This paper presents a research strategy for evaluating the capability of constructed, restored, and natural wetlands to assimilate and process pesticides associated with agricultural runoff from croplands. A modeling approach that is central to this research strategy is presented and the mathematical foundation is explicitly stated. This approach generates predictions that can be experimentally and rigorously tested. Criteria for selection of "model" pesticides for experimentation include factors such as use patterns and amounts as well as intrinsic characteristics of the pesticide. The design of the experimental constructed wetland cells for this research includes water flow and depth control, clay liners to prevent infiltration, and wetland vegetation as a variable. The experimental strategy should permit optimal transfer of study results from site to site and ultimately provide recommendations for pesticides that are compatible with wetlands as well as design characteristics for constructed wetlands to be used with specific crop-pesticide combinations.

COMPARISON OF WATER QUALITY FUNCTIONS OF CREATED, RESTORED AND NATURAL WETLANDS

Wetlands have numerous functions and values including water quality improvement (Greeson et al., 1979; Mitsch and Gosselink, 1986; Hammer and Bastian, 1989). Protection, creation, and restoration of wetlands may be of great value in integrated strategies for controlling nonpoint source (NPS) runoff. As interfaces between cropland and water, some wetlands are subject to episodic NPS agricultural runoff (Wauchope, 1978). These wetlands can retain and process many NPS contaminants (Cooper, 1989). When contaminants such as pesticides are processed in wetlands, their impacts are not subsequently realized in downstream lakes, rivers, streams, and reservoirs.

NPS agricultural contaminants entering wetlands may include physical, chemical, or biological contaminants (Wauchope, 1978; Gersberg et al., 1987). Physical contaminants include materials such as particulates or detritus. Chemical contaminants include pesticides, organics, and inorganics such as nutrient salts. Biological contaminants include bacteria from animals such as dairy cattle and chickens, as well as protozoa and other pathogens (Portier and Palmer, 1989). Probably the most important NPS agricultural contaminants causing problems in receiving systems are particulates, nutrients, and pesticides. (Wauchope, 1978; Wallach, 1991). Frequently, pesticides that are washed from croplands by rainfall and cause adverse impacts on adjacent waters come from fields where a wetland buffer strip was not maintained. In these instances, water quality problems and even fish kills may be a consequence of the loss of the wetland (Wolverton and Harrison, 1975).

The objective of this paper is to present, by example, a strategy for developing design criteria for wetlands to control NPS runoff from agricultural systems with emphasis on pesticides. This research strategy examines factors affecting the limits of wetlands for assimilating agricultural pesticides. A long-term goal is to couple the appropriate pesticides with correctly designed constructed or restored wetlands for maximal retention and processing of NPS pesticides.

RESEARCH FOUNDATION AND APPROACH

Farmers in the southeastern United States have developed for agricultural use most of the available land bordering rivers, streams, reservoirs, and lakes. The notion of "plowing with one wheel of the tractor in water" became popular in order to maximally utilize the rich flood plain soils. These former wetland areas have been lost, particularly in the Mississippi Valley, at an alarming rate (Horwitz, 1978). The research described in this paper is based on the premise that wetlands' functions and values ranging from their chemical and material processing ability to their values for wildlife propagation can be coupled to mitigate the effects of NPS agricultural runoff, particularly pesticide components. Since wetlands have a unique ability to retain and process materials, it seems reasonable that constructed or restored wetlands as buffer strips between agricultural activities and receiving aquatic systems could mitigate impacts of pesticides in this runoff.

The factors influencing pesticide fate and effects in agricultural systems and wetlands have been a matter of fairly recent investigation (Wauchope, 1978; Reinert and Rodgers, 1984). We currently have a relatively good understanding of the characteristics of aquatic systems as well as the characteristics of pesticides that regulate their fate and effects. Modeling of pesticide fate in agricultural systems and adjacent wetlands yields predictions of pesticide runoff in aqueous and particulate form, biotransformation rates, wetland retention, photolysis rates, hydrolysis rates, sorption, and other factors that may influence pesticide fate (Reinert and Rodgers, 1984). This knowledge can be used to enhance or select for factors in wetlands that promote processing of pesticides (Table 1). The research described in this paper consists of four parts. First, a modeling effort was undertaken, focusing on characteristics of pesticides and wetlands that would be important in agricultural pesticide retention and transformation. Next, pesticides were selected for testing and evaluation. Third, experimental wetlands were designed and constructed to test hypotheses regarding the fate of pesticides in wetlands. And fourth, hypotheses are being tested and conclusions drawn.

MODELING EFFORT

The pesticide transfer and transformation model used to guide this research is based on wetland physical, chemical and biological characteristics and processes that regulate the fate and persistence of a pesticide (Table 1). The rate of removal of a pesticide in a wetland is compared with the pesticide's residence time (PRT) in the wetland to determine the propensity of a wetland to reduce the concentration of a pesticide in downstream aquatic systems and thereby to mitigate subsequent nontarget effects. For this model, the PRT is defined as the time available in the wetland for pesticide transfer and occurrence of transformation processes such as volatilization, hydrolysis, photolysis, and biotransformation (Reinert and Rodgers, 1984). The PRT is a function of: (1) runoff events that add more pesticide to the wetland, or (2) processes that transfer the pesticide out of the wetland. Although this is a simplified view of pesticide transport (Wauchope, 1978), it will suffice until we have sufficient data from wetlands to permit more sophisticated or complicated analysis.

The pesticide transfer and transformation model used for these wetland studies is a first-order or pseudo first-order mass balance model in which the pesticide transfer and transformation rate is directly proportional to the pesticide concentration. In this model, the overall transformation and transfer rate coefficient (K_T) is an aggregate of the individual rate coefficients for the processes noted above (e.g., hydrolysis, biotransformation, etc.). The implication of the model is that pesticide removal processes in wetlands can be adequately described by simple exponential decay, and it perhaps implies a simpler mechanism than actually occurs. As fate and persistence data are accumulated for pesticides in wetlands, the frequency of deviations from simple exponential behavior can be evaluated. However, it should be noted that similar modeling efforts have been very useful and provided predictive capabilities for wastewater processing in wetland situations (Bavor et al., 1989; Kadlec, 1989; Steiner and Freeman, 1989). To date, the mass balance approach with exponential modeling has been accurate for a variety of pesticides and wetlands (Reinert and Rodgers, 1984; Reinert et al., 1988; Cassidy and Rodgers, 1988).

The first-order or pseudo first-order model of pesticide fate (transfer and transformation) in a wetland is the following:

$$C_t = C_i\, \theta^{-K_T t} \qquad (1)$$

where C_t and C_i are pesticide concentration at time t and the initial concentration, respectively, and where θ is the specific base of the decay function. For this exponential model, K_T is the overall transfer and transformation rate coefficient with units of reciprocal time. A useful expression of K_T is the pesticide transformation and transfer half-life ($T_{1/2}$)

which is the time required to reduce the mass of a pesticide to 50 percent of its original value. Mathematically:

$$T_{½} = \frac{\ln(2)}{K_T} = \frac{0.693}{K_T} \qquad (2)$$

For experimental and design purposes relative to this study, we can express the model for pesticide transfer and transformation in wetlands as a basic equation:

$$\frac{C_T}{C_i} = \theta^{-0.693\,(PRT/T_{½})} \qquad (3)$$

C_T / C_i is the removal ratio for a pesticide at the end of a given PRT.

Explicitly stated, the assumptions behind the pesticide transfer and transformation model are: (1) pesticide transfer and transformation in wetlands follows first-order or pseudo first-order kinetics, and (2) transport of a pesticide in a wetland can be reasonably approximated by a single number, the PRT. The PRT in a given wetland will be determined by the characteristics of that wetland such as plant density, porosity, wetland dimensions, water flow, pesticide retention factors such as sorption, etc., as well as runoff events that influence pesticide inputs to the wetlands. The residence time of a pesticide in a wetland is also a function of the intrinsic chemical character of the pesticide. As noted above, these assumptions will probably permit accurate predictions for most situations until more data are available to indicate more complicated or appropriate approaches.

Predictions from the pesticide transfer and transformation model are summarized in Figures 1 and 2. In Figure 1, the $T_{½}$ is related to the PRT and pesticide removal in wetlands is estimated. For example, to achieve at least 95 percent removal of a pesticide in a wetland with a PRT of 1 month, the critical $T_{½}$ is approximately 7 d. On the other hand, if the PRT is 6 months, the $T_{½}$ required for 95 percent removal may be considerably longer, on the order of 50 d. If the $T_{½}$ of a pesticide and the PRT in a wetland are 15 and 50 d, respectively, then one would expect removal of at least 90 percent. When these pesticide removal rates are compared to environmental toxicology information, the mitigation capabilities of wetlands for reducing effects of pesticides on downstream aquatic systems can be predicted.

The modeled relationship of pesticide removal with respect to $T_{1/2}$ and PRT is illustrated in Figure 2. This figure shows the amount of pesticide remaining as a function of the $T_{1/2}$/PRT ratio. From this illustration, it is clear that significant pesticide removal ($\leq$50 percent remaining) will occur when the $T_{1/2}$ equals the PRT ($T_{1/2}$/PRT = 1). Pesticide removal increases still further (or the amount of pesticide remaining decreases) as the $T_{1/2}$ becomes a smaller fraction of the PRT ($T_{1/2}$/PRT < 1). At $T_{1/2}$/PRT ratios of 3.0 or less, removal is essentially linear. A further prediction from the model that can be derived from Figure 2 is that the utility of wetlands for pesticide removal diminishes rapidly when the $T_{1/2}$ exceeds the PRT. At $T_{1/2}$/PRT ratios greater than 1, the change in $T_{1/2}$ required to significantly reduce the amount of pesticide remaining in a wetland is large relative to that required when the $T_{1/2}$/PRT is less than 1. For example, at a $T_{1/2}$/PRT ratio of 8, less than 10 percent of the pesticide will be removed in the wetland. Pesticides with these $T_{1/2}$/PRT ratios will not be practically removed by wetlands. Another key practical observation from this modeling effort is that pesticide removal efficiency in different types of wetlands may vary strongly with factors that influence residence times in differing wetlands and physical, chemical and biological factors that regulate pesticide half-lives. In addition, it was apparent from the modeling effort that one could predict the physical dimensions of a wetland necessary to obtain a required pesticide removal fraction if the relationships between PRT and wetland characteristics such as vegetation density and hydrosoil characters are known. Clearly, this would be a profitable area for further study.

The mass balance approach for pesticide fate is used in this research strategy to evaluate the role of wetlands in mitigating the effects of pesticides in NPS runoff. During the initial sensitivity analysis of the model, it was apparent that processing of the pesticide would be relative to the retention of the pesticide in the wetland. Retention of the pesticide in the wetlands was a function of sorption and contact time in the wetland. Sorption is largely a function of the available mass of plants or vegetation in the wetland as well as the hydrosoil character (Reinert and Rodgers, 1987). Contact time that influenced retention of the pesticide in the wetland was largely a function of volume and flow. Processing of the pesticide in the wetland was a function of biotransformation, photolysis, hydrolysis, and other biological and chemical processes. The wetland dimensions required to optimize retention and processing could be discerned from the relationship between the half-life of the pesticide and the wetland physical/chemical character. This modeling effort yielded a series of hypotheses that could be experimentally tested in replicated constructed wetlands cells using some "model" or example pesticides.

SELECTION OF PESTICIDES FOR TESTING

Pesticide selection for experimental purposes was based on three factors. First, the use pattern and volume of pesticide used in Mississippi were selection criteria. It was considered important to investigate pesticides that are heavily used in large volumes and widely used for a number of crops. Secondly, the toxicology and fate characteristics of the pesticide were important. Initially, we wanted to choose a pesticide that would theoretically be very efficiently removed and processed in a wetland, and then choose a pesticide that we expected would be largely incompatible with wetlands in order to illustrate the bounds of wetland processing abilities. As an additional criterion, we wanted to choose a pesticide for which we had analytical and toxicological experience. We need to have efficacious analytical techniques for the pesticide in water, sediment, and plant and animal material. If we have a toxicological data base, animals and plants may be used to evaluate the bioavailability of the pesticide as it is retained in the wetland.

Based on available data (Wauchope, 1978; Reinert and Rodgers, 1987; State of Mississippi, 1990), a number of candidate pesticides were evaluated (Table 2). In Mississippi, both insecticides and herbicides are widely used in relatively large volumes on soybeans, cotton, wheat, and rice (State of Mississippi, 1990). Pyrethroid insecticides are the predominant materials used and are viable candidates for use with wetland buffer strips. Pyrethroids are not particularly water soluble (solubility, in water 10-20 ppb) and readily sorb to plants and soil. They are generally subject to hydrolysis, photolysis, and biotransformation. Herbicides that are widely used include triazines, fluridone, and copper-based materials (State of Mississippi, 1990). These herbicides may not be compatible with wetland buffer strips if they cause direct impacts on the wetland vegetation.

It is apparent that the timing and magnitude of the rainfall event after pesticide application is crucial in determining pesticide loss in runoff. Measured losses of pesticides in runoff from agricultural fields ranges from 0.5 to 5 percent, depending on weather and slope of the field. In terms of ecological impacts on receiving aquatic systems, the mass of pesticide lost from an agricultural field may not be the primary issue. Of primary concern may be the intermediate intensity rainfall events that yield maximal concentrations of pesticide in receiving aquatic systems. Small rainfall events may not yield sufficient mass of pesticide residue when diluted by receiving systems to be a problem. Large or catastrophic rainfall events (which produce runoff loses of 2 percent or more of the applied amount of pesticide; Wauchope, 1978) may actually result in lower concentrations of pesticides in receiving aquatic systems than intermediate events (Bailey et al., 1974). Thus, impacts of pesticides in runoff from agricultural fields may be most severe from rainfall events that are large enough to mobilize pesticides but small enough to avoid excessive dilution resulting in maximum concentrations in the receiving system (Trichell et al., 1968; Wauchope, 1978; Wallach, 1991). All of the pesticide may be mobilized if the rainfall occurs soon after application. It is important to recall that pesticide concentration in runoff may vary more than an order of magnitude during a single event. The most precipitous impacts on non-target species in receiving aquatic systems are usually related to the maximum exposure concentration and not the average of a runoff event. The fraction of pesticide mass sorbed to particulates and the dissolved or aqueous fraction that is transported depends on the character of the pesticide and the sorbents or the agricultural soil (Rodgers et al., 1987). Still at issue is the bioavailability or bioactivity of particulate-bound and dissolved pesticides in agricultural runoff. This is an issue that can be, at least partially, resolved by the research strategy.

EXPERIMENTAL DESIGN

The experimental design used for testing model-derived hypotheses incorporates two major variables. First, the major biological component, the vegetation, of the wetland design was crucial. Wetland plants have differing growth habits and physiology that affect their ability to perform in buffer strips to process agricultural pesticide runoff (Wolverton and Harrison, 1975; Gersberg et al., 1985; Bowmer, 1987; Good and Patrick, 1987; Stengel and Schultz-Hock, 1989). For example, wetland plants may contribute to or promote formation of anaerobic, low-redox hydrosoils or they may aerate their root zone forming a largely aerobic, high-redox potential hydrosol (MacMannon and Crawford, 1971; Whitlow and Harris, 1979; Mendelssohn et al., 1981; Faulkner and Richardson, 1989; Guntenspergen et al., 1989).

For this research, three wetland plants were chosen with widely varying physiology and growth habits. *Typha latifolia (cattail)* will produce aerobic zones and considerable biomass in the wetland hydrosoil (Rodgers et al., 1983; Faulkner and Richardson, 1989). *Scirpus cyperinus* (bulrush) and *Zizania aquatica* (wild rice) should permit formation of anaerobic zones in the wetland hydrosoils (Whigham and Simpson, 1977; Whitlow and Harris, 1979). The influence of a wetland plant on the hydrosoil character as well as providing sorption surface for pesticide retention were predicted by the modeling effort to be important factors in pesticide removal efficiency in the experimental wetland cells.

Secondly, the physical and chemical character of the wetland was also very important for this research strategy. The dimensions of the wetland are important in retention capacity for runoff from rainfall events (Figure 3). These experimental wetland cells are constructed with water recirculation capability so a variety of runoff scenarios can be simulated (Kadlec, 1989; Steiner and Freeman, 1989). Water depth in each wetland cell can be readily controlled to provide a suitable habitat for the chosen vegetation. Each wetland is lined below the hydrosoil with 10 cm of compressed bentonite clay with permeability less than 10^{-6}cm/sec to prevent infiltration. Ground water monitoring wells are located at the periphery of the constructed wetland experimental area to ensure that the pesticides do not penetrate the clay liners. The length is extended on some of the constructed wetland cells to permit longer retention times at a given flow relative to adjacent, shorter wetland cells (Figure 3). The wetland experimental area is surrounded by a subsurface drain field and diversion ditch to ensure that no unwanted surface or ground water enters. A catch basin is located downstream for emergency capacity to accept water from the experimental wetland cells. The Wetland Research area is located on the 300-ha Biological Field Station of the University of Mississippi near Oxford (Lafayette County).

TESTABLE HYPOTHESES-DISCUSSION

Using the research strategy presented, a series of experiments are underway that we have designed to test important hypotheses regarding the ability of wetlands to mitigate the impacts of pesticides in agricultural runoff. These hypotheses are presented as questions below.

- Does the type of wetland vegetation play an important role in the efficiency of pesticide retention and processing in wetlands?

- Can the relatively simple transfer and transformation model used to predict pesticide fate in wetlands be validated or modified to obtain satisfactory estimates?
- Are insecticides processed by wetlands more efficiently than herbicides?
- Do constructed, essentially monospecific wetlands perform better than successional or multi-species wetlands?
- What are the maintenance costs for wetland buffer strips? Can the wetlands function efficiently for a period of time without considerable maintenance?
- Does sediment or particulate loading influence wetland functions? Are sorbed pesticides bioavailable or bioactive?
- Do the wetland sediments or hydrosoils regulate the physiology of the vegetation or are the sediment or hydrosoil characteristics regulated by the extant vegetation?
- Can wetland buffer strips and pesticides be compatibly linked to practical recommendations that can be incorporated into routine agricultural practices?
- Are wetland functions or values compromised when wetlands are used as buffer strips for nonpoint source pesticide runoff from agricultural croplands?

These and other ancillary hypotheses will require several years of experimental efforts to resolve. It is hoped that the experimental strategy will permit an optimal effort to gather information that will be readily transferable from site to site and of maximum utility.

REFERENCES

Bailey, G. W., A. R. Savank, Jr., and H. P. Nicholson, 1974. Predicting pesticide runoff from agricultural land: a conceptual model. Journal of Environmental Quality, 3: 95-102.

Bavor, H. J., D. J. Roser, P. J. Fisher, and I. C. Smells, 1989. Performance of solid-matrix wetland systems viewed as fixed film bioreactors. pp. 646-657. In: D. A. Hammer (ed.), Constructed Wetlands for Wastewater Treatment. Lewis Publishers, Inc., Chelsea, MI.

Bowmer, K. H, 1987. Nutrient removal from effluents by an artificial wetland: influence of rhizosphere aeration and preferential flow studied using bromide and dye tracers. Water Research, 21: 591-599.

Cassidy, and J. H. Rodgers, Jr., 1988. Response of Hydrilla (*Hydrilla verticillata*) (L.f.) Royle) to Diquat and a model of uptake under nonequilibrium conditions. Environmental Toxicology and Chemistry, 8: 133-140.

Cooper, C. M., 1989. Status of current technology on constructed wetlands. Submitted to the DEC Task Force, National Sedimentation Laboratory, USDA-ARS, Oxford, MS.

Faulkner, S. P. and C. J. Richardson, 1989. Physical and chemical characteristics of freshwater wetland soils. pp. 41-72. In: D. A. Hammer (ed.), Constructed Wetlands for Wastewater Treatment. Lewis Publishers, Inc., Chelsea, MI.

Gersberg, R. M., R. Bermer, S. R. Lyon, and B. V. Elkins, 1987. Survival of bacteria and viruses in municipal wastewater applied to artificial wetlands. pp. 237-245. In: K. R. Reddy and W. H. Smith (eds.), Aquatic Plants for Water Treatment and Resource Recovery. Magnolia Publishing, Orlando, FL.

Gersberg, R. M., B. V. Elkins, S. R. Lyons, and C. R. Goldman, 1985. Role of aquatic plants in wastewater treatment by artificial wetlands. Water Research, 20: 363-367.

Good, B. J. and W. H. Patrick, Jr., 1987. Root-water-sediment interface process. pp. 359-371. In: K. R. Reddy and W. H. Smith (eds.), Aquatic Plants for Water Treatment and Resource Recovery. Magnolia Publishing, Orlando, FL.

Greeson, P. E., J. R. Clark, and J. E. Clark (eds.), 1979. Wetland Functions and Values: The State of our Understanding. American Water Resources Association, Minneapolis, MN.

Guntenspergen, G. R., F. Sterns, and J. A. Kadlec, 1989. Wetland vegetation. pp. 89-103. In: D. A. Hammer (ed.), Constructed Wetlands for Wastewater Treatment. Lewis Publishers, Inc., Chelsea, MI.

Hammer, D. A. and R. A. Bastian, 1989. Wetlands ecosystems: natural water purifiers. pp. 5-19. In: D. A. Hammer (ed.), Constructed Wetlands for Wastewater Treatment. Lewis Publishers, Inc., Chelsea, MI.

Horwitz, E. L., 1978. Our Nation's Wetlands – An Interagency Task Force Report. Council on Environmental Quality, U.S. Government Printing Office, Washington, DC.

Kadlec, R. H., 1989. Hydrologic factors in wetland water treatment. pp. 21-40. In: D. A. Hammer (ed.), Constructed Wetlands for Wastewater Treatment. Lewis Publishers, Inc., Chelsea, MI.

MacMannon, M. and R. M. M. Crawford, 1971. A metabolic theory of flooding tolerance: the significance of enzyme distribution and behavior. New Phytologist, 10: 299-306.

Mendelssohn, I. A., K. L. McKee, and W. H. Patrick, Jr., 1981. Oxygen deficiency in *Spartina alterniflora* roots: metabolic adaptation to anoxia. Science, 214: 439-441.

Mitsch, W. J. and J. G. Gosselink, 1986, Wetlands. Van Nostrand Reinhold, New York, NY.

Portier, R. J. and S. J. Palmer, 1989. Wetlands microbiology: form, function, process. pp. 21-40. In: D. A. Hammer (ed.), Constructed Wetlands for Wastewater Treatment. Lewis Publishers, Inc., Chelsea, MI.

Reinert, K. H., M. L. Hinman, and J. H. Rodgers, Jr., 1988. Fate of Endothall during the Pat Mayse Lake (Texas) aquatic plant management program. Archives of Environmental Contamination and Toxicology, 17: 195-199.

Reinert, K. H. and J. H. Rodgers, Jr., 1987. Fate and persistence of aquatic herbicides. Reviews of Environmental Contamination and Toxicology, 98: 61-98.

Reinert, K. H. and J. H. Rodgers, Jr., 1984. Validation trial of predictive fate models using an aquatic herbicide (Endothall). Environmental Toxicology and Chemistry, 5: 449-461.

Rodgers, J. H., Jr., K. L. Dickson, F. Y. Salem, and C. A. Staples, 1987. Bioavailability of sediment-bound chemicals to aquatic organisms – some theory, evidence and research needs. pp. 245-266. In: K. L. Dickson, A. W. Maki, and W. A. Brungs (eds.), Fate and Effects of Sediment Bound Chemicals in Aquatic Systems. Pergamon Press, Elmsford, NY.

Rodgers, J. H., Jr., M. E. McKevitt, D. O. Hammerland, K. L. Dickson, and J. Cairns, Jr., 1983. Primary production and decomposition of submergent and emergent aquatic plants of two Appalachian rivers. pp. 298-301. In: T. D. Fontaine III and S. M. Bartell (eds.), Dynamics of Lotic Ecosystems. Ann Arbor Science Publishers, Ann Arbor, MI.

State of Mississippi, 1990. Pesticide Use. Department of Agriculture and Commerce, Division of Plant Industry, Mississippi State, MS.

Steiner, G. R. and R. J. Freeman, Jr., 1989. Configuration and substrate design considerations for constructed wetlands wastewater treatment. pp. 363-377. In: D. A. Hammer (ed.), Constructed Wetlands for Wastewater Treatment. Lewis Publishers, Inc., Chelsea, MI.

Stengel, E. and R. Schultz-Hock, 1989. Denitrification in artificial wetlands. pp. 484-491. In: D. A. Hammer (ed.), Constructed Wetlands for Wastewater Treatment. Lewis Publishers, Inc., Chelsea, MI.

Trichell, D. W., H. L. Morton, and M. G. Merkle, 1968. Loss of herbicides in runoff water. Weed Science, 16: 447-449.

Wallach, R., 1991. Runoff contamination by soil chemicals: time scale approach. Water Resources Research, 27: 215-223.

Wauchope, R. D., 1978. The pesticide content of surface water draining from agricultural fields – a review. Journal of Environmental Quality, 7: 459-472.

Whigham, D. F. and R. L. Simpson, 1977. Growth, mortality, and biomass partitioning in freshwater tidal wetland populations of wild rice (*Zizania aquatica* var. *aquatica*). Torry Botanical Club Bulletin, 104: 347-351.

Whitlow, T. H. and R. W. Harris, 1979. Flood tolerance in plants: a state-of-the-art review. Technical Report E 79-2. U.S. Army Engineers Waterways Experiment Station, Vicksburg, MS.

Wolverton, B. C. and D. D. Harrison, 1975. Aquatic plants for removal of mevinphos from the aquatic environment. Journal of the Mississippi Academy of Sciences, 19: 84-88.

Table 1. Transfer and removal processes in wetlands that are important in mitigation of nonpoint source pesticide runoff.

TRANSFER PROCESSES	REMOVAL PROCESSES
flow	volatilization
sorption	photolysis
solubility	hydrolysis
retention	biotransformation
infiltration	

Table 2. Pesticides commonly used in Mississippi (State of Mississippi, 1990).

Tralomethrin (Scout)
Permethrin (Ambush/Pounce)
Cypermethrin (Ammo/Cymbush)
Flucythrinate (Payoff)
Fenvalerate (Pydrin)
Carbaryl (Sevin)
Malathion (Malathion 5 EC)
Trifluralin (Treflan)
Dimilin (Scepter)
Metrubuzine (Sencor)
Benomyl (Dupont Benlate)
Mancozeb (Manzate 200)

LIST OF FIGURES

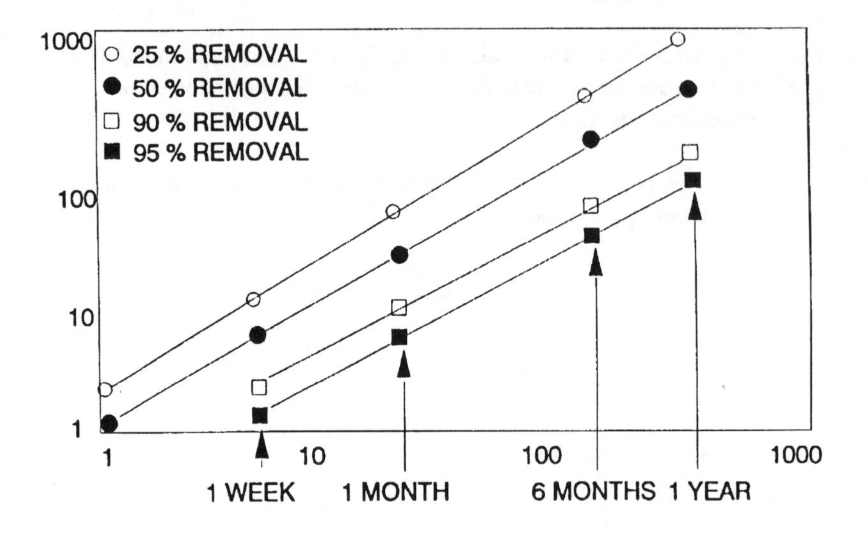
TRANSFER AND TRANSFORMATION HALF-LIFE NEEDED
TO ACHIEVE NOTED PRESTICIDE REMOVAL (DAYS)
25 % REMOVAL
50 % REMOVAL
90 % REMOVAL
95 % REMOVAL
1000
100
10
1
1
10
100
1000
1 WEEK
1 MONTH
6 MONTHS
1 YEAR
PESTICIDE RESIDENCE TIME (DAYS)

Figure 1

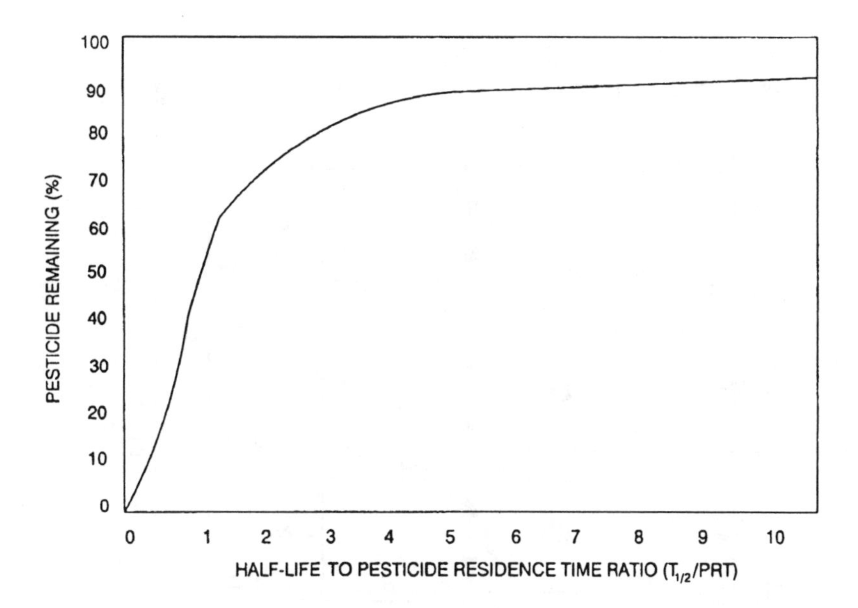
100
90
80
70
60
50
40
30
20
10
0
PESTICIDE REMAINING (%)
0
1
2
3
4
5
6
7
8
9
10
HALF-LIFE TO PESTICIDE RESIDENCE TIME RATIO ($T_{1/2}$/PRT)

Figure 2

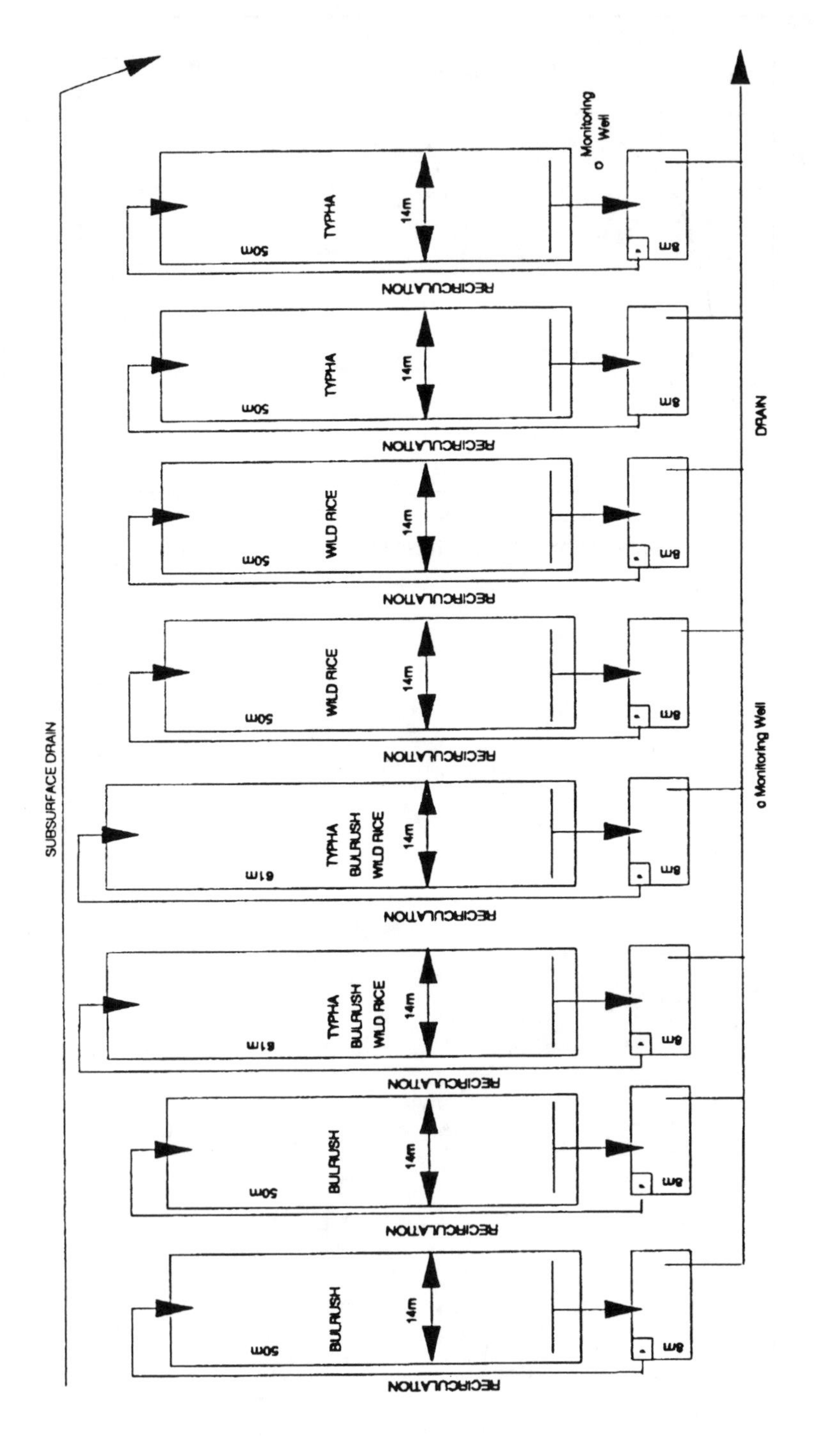
SUBSURFACE DRAIN
DRAIN
o Monitoring Well
o Monitoring Well
TYPHA
50m
14m
8m
RECIRCULATION
TYPHA
50m
14m
8m
RECIRCULATION
WILD RICE
50m
14m
8m
RECIRCULATION
WILD RICE
50m
14m
8m
RECIRCULATION
TYPHA
BULRUSH
WILD RICE
61m
14m
8m
RECIRCULATION
TYPHA
BULRUSH
WILD RICE
61m
14m
8m
RECIRCULATION
BULRUSH
50m
14m
8m
RECIRCULATION
BULRUSH
50m
14m
8m
RECIRCULATION

Figure 3

CHAPTER 6

Ancillary Benefits and Potential Problems with the use of Wetlands for Nonpoint Source Pollution Control

Robert L. Knight

ABSTRACT

Wetlands utilized for the control of nonpoint source (NPS) pollution provide a variety of secondary benefits in addition to their primary roles of flood attenuation and water quality enhancement. Ancillary benefits provided by these wetlands typically include photosynthetic production, secondary production of fauna, food chain and habitat diversity, export to adjacent systems, and services to human society such as aesthetics, hunting, recreation, and research. While most NPS wetlands provide these ancillary benefits, the quantitative magnitude of these functions may vary greatly from one system to another. Similarly, some functions provided by NPS control wetlands may be detrimental to the wetland flora and fauna or to society. This paper provides a brief review of the ecological knowledge available about these functions and provides guidance on optimizing the appropriate ancillary benefits and avoiding undesirable side effects while achieving primary NPS control goals.

INTRODUCTION

Wetlands perform a variety of functions including physical, chemical, and biological processes that create economic or aesthetic values to society or life-support for plant and animal populations (Gosselink and Turner, 1978; Sather and Smith, 1984; Mitsch and Gosselink, 1986; Erwin, 1990; and Kusler and Kentula, 1990). A partial list of wetland functions includes water storage/flood attenuation, nutrient assimilation/transformation, sediment storage, photosynthetic production, secondary production, food chain and habitat diversity, export of carbon and organisms to adjacent ecosystems, and aesthetic/recreational/educational human uses.

While a specific wetland may perform some or all of these functions, the relative magnitude of each function exhibited by a wetland, as well as the number of functions exhibited, is highly variable. The observation that the term "wetland" includes a diverse array of different ecosystems, with a diversity of abiotic and biotic forcing functions, leads to the deduction that "all wetlands are not created equal," Published summaries of wetland function display a broad range of quantitative functional attributes, even for a single wetland plant community type (Mitsch and Gosselink, 1986). For example, net primary productivity (NPP) varies from as low as 50 $g/m^2/y$ in arctic tundra to 3,500 $g/m^2/y$ in southeastern marshes, and litter production varies from 460 to 2,000 $g/m^2/y$ in freshwater marshes (Nixon and Lee, 1986). Total wetland nitrogen assimilation rate varies from less than 0.08 to over 90 $g/m^2/yr$ (Nixon and Lee, 1986; Knight, 1990). While the functional attributes of wetlands are variable in quality and quantity, approximate functional levels for individual wetlands can be estimated based on a knowledge of the structure of the wetland and the array of forcing functions affecting the wetland (Erwin, 1990). In the example above, freshwater marshes display a wide range of NPP rates in North America, yet a knowledge of a specific marsh's latitude, hydrology, vegetation, and soil type can greatly increase the accuracy of an estimate of annual NPP.

The rapidly expanding understanding of the relationships between a wetland's structure and function can be used to predict and enhance the functions of wetlands designed for control of nonpoint source (NPS) pollutants. While wetlands can be used to accomplish the primary objectives of NPS control (reduction of peak stormwater flows and control of suspended and dissolved pollutants), ancillary benefits also may be achieved through thoughtful site selection and design. This paper provides a summary of the ancillary benefits and potential problems associated with purposely including wetlands in NPS control strategies, and recommends specific design features that can be used to optimize this technology. Other papers in this volume (Mitsch; Hammer) and elsewhere (Hammer, 1989) discuss the primary objectives of NPS control with wetlands. Research concerning wetland evaluation techniques (Golet, 1978; Greeson et al., 1979; Richardson, 1981; Kusler and Riexinger, 1986) provides a yardstick for comparing the ancillary benefits of wetlands constructed for NPS pollution control with natural wetland functions.

ANCILLARY BENEFITS OF NPS CONTROL WETLANDS

The primary objectives of most NPS control projects utilizing wetlands include (1) water storage/flood attenuation and (2) water quality enhancement through assimilation/transformation of sediments, nutrients, and toxic chemicals. The ancillary or secondary benefits that may be gained from NPS control wetlands include (1) photosynthetic production, (2) secondary production of fauna, (3) food chain and habitat diversity, (4) export to adjacent ecosystems, and (5) aesthetic/recreational/educational human uses. The potential for inclusion of each of these ancillary benefits in NPS control wetlands as well as their potential quantitative functional levels are described below.

Photosynthetic Production

One function shared by all wetlands is photosynthetic production by vascular and non-vascular plants. The magnitude of this primary productivity, however, varies greatly between wetland types and even within a single wetland habitat type, due to varying environmental forcing functions at work on individual wetland areas. Wetlands generally have higher NPP than adjacent uplands (900 to 2,700 $g/m^2/y$ in wetlands compared with approximately 500 $g/m^2/y$ in grasslands) due to the subsidies of water and nutrients from these adjacent systems (Richardson, 1978).

Enhancing the primary production of wetlands constructed for NPS control may or may not be desirable depending on the goals of the project. If a goal is to contribute organic matter as the basis of a food chain leading to domesticated or wild animal populations, then factors that may limit primary productivity can be supplied to some extent through system design and operational control. Forcing functions most frequently limiting primary productivity of constructed wetlands include light, water, macronutrients, and micronutrients. In emergent marshes, light is the most likely factor to limit algal production. If a project goal is reduction of algal suspended solids, a typically included design feature is a densely vegetated emergent zone at the downstream end of the wetland treatment system. If algal productivity is desired to enhance an aquatic food chain (for example, fish or shellfish culture), open water and deeper areas should be included in the design.

In constructing wetlands dependent upon stormwater, water itself may be the most limiting environmental factor during extended dry periods. High primary production is not realized in dry wetlands. To maintain a high level of primary productivity in a constructed wetland during all seasons, an alternate water source should be provided. Highest NPP is generally measured in shallow (less than 0.3 m), regularly flooded emergent marshes (Brown et al., 1979). This may be due to the availability of water combined with higher sediment dissolved oxygen levels in shallow or flowing systems (Gosselink and Turner, 1978). More natural, fluctuating water levels will generally result in lower NPP. Constructed wetlands designed to simulate natural wetland hydroperiods need to include water-level control structures to prevent damaging flood flows and to replicate the slow bleed-down of water levels following storms (Livingston, 1989). Prolonged high water in a constructed wetland will result in a rapid successional change from emergent, shallow water vegetation to an aquatic system dominated by phytoplankton, filamentous algae, or floating aquatics (Guntenspergen et al., 1989). If these prolonged flood events occur on several occasions each year, with dry spells interspersed, a highly stressed and potentially unproductive plant community will result.

Faunal Production

Secondary production of herbivorous and/or carnivorous animals in wetlands constructed for NPS control may be directed to aquaculture of "domesticated" species or to enhancement of wildlife populations. Generally, greater operational control is required to direct secondary production to "domesticated" species such as crayfish, fish, or other forms of aquaculture. Wengrzynek and Terrell (1990) have developed a generic constructed wetland design for NPS control that may incorporate baitfish or freshwater mussel production as an ancillary benefit. Wetlands dominated by grasses also may be highly productive areas for livestock grazing.

To many conservationists, the most exciting ancillary benefit of the construction of wetlands for NPS control is enhancement of wildlife populations. Wetland and impoundment design for wildlife enhancement is relatively well known (Weller, 1978; Smith et al., 1989; Weller, 1990). A diversity of plant species and growth forms with a variety of plant and seed maturity dates will provide a wider range of wildlife niches within the wetland. Total wildlife production also may be most closely correlated with water quantity and quality. Wildlife are attracted to wetlands that have perennial water, and areas that are flooded less frequently generally will have smaller populations of wetland-dependent wildlife species. However, fluctuating water levels create additional niches and result in higher wildlife diversity. Wetlands created for water quality enhancement also may provide expanded habitat for threatened or endangered wildlife species.

Water quality is important to wildlife production through its control on primary production. Constructed wetlands receiving waters with higher nutrient content generally have larger wildlife populations. Increased NPP is transferred primarily through the detritus food chain to invertebrates and small fish, reptiles, amphibians, and birds. As long as these intermediate consumer populations are not restricted from colonizing and increasing their numbers within a wetland with high plant productivity, they will develop a food base for the more highly visible avifauna typical of constructed wetlands.

If inflow waters have lower nutrient levels, typical of nonpoint sources from less developed watersheds, then food chain support will be less than the maximum possible level. In a lower productivity NPS pollution constructed wetland, wildlife species diversity may be higher than in a wetland with higher primary productivity. Typically, a wetland's value for wildlife increases with the association of neighboring undeveloped upland habitat.

The physical design features of a wetland may have a greater influence than nutrient levels on faunal diversity and abundance. For example, waterfowl populations are enhanced by the provision of open water areas interspersed with deep emergent marsh and upland islands. An approximate ratio of wetland area devoted to marsh and open water that will provide maximum habitat for a variety of waterfowl is about 1:1 (Weller, 1978).

Wading birds require a different habitat mix than waterfowl. These species require shallow, sparsely vegetated littoral areas or perching substrates adjacent to open water areas. NPS constructed wetlands can be designed to provide a broad shelf of emergent marsh with water depths of less than 20 to 30 cm to benefit populations of wading birds. Deep, open water areas adjacent to a shallow marsh provide additional foraging habitat for wading birds. Open water areas and the transitional ecotones between marsh, open water areas, and adjacent uplands help promote higher populations of herptiles and fish which, in turn, provide a forage base for wading and diving birds. Inclusion of a diverse fish population during system startup or through natural immigration will lengthen the consumer food chain and provide potential support for raptors such as ospreys, hawks, eagles, and kites.

Many other varieties of birds will colonize constructed wetlands, depending on regional occurrence and available habitats. If living or dead trees are included in the wetland, they will serve as perching and possible nesting sites for numerous bird species (Hair et al., 1978). Nesting boxes may be provided to encourage use of the site by wood ducks and owls. Upland islands surrounded by open water provide protection for ground-nesting bird species.

A variety of small or large mammals may utilize constructed wetlands, depending on available food and habitat resources. Small mammals will develop large populations on upland areas adjacent to and within the constructed wetlands. These populations provide a forage base for raptors and large wading birds. Larger mammals also may be included, which adds to overall system diversity and creates the possibility of a byproduct of animal skins. Nutria, muskrats, and beaver generally will not colonize a constructed wetland unless perennial water is available. Large mammals may provide important feedback control of NPP in wetlands due to their ability to rapidly reduce biomass for lodges and food and to maintain patches of high net productivity, early successional marsh (Weller, 1978; Hair et al., 1978).

Human Uses

A major functional value of wetlands is their importance for consumptive (plant harvesting, livestock grazing, hunting, aquaculture, etc.) and nonconsumptive (aesthetics, recreation, and research) human uses (Nash, 1978; Reimold and Hardisky, 1978; Sather and Smith, 1984; and Smardon, 1988).

Consumptive uses such as waterfowl hunting and fur trapping are more easily quantified than nonconsumptive functions (Chabreck, 1979). Nonconsumptive human uses of wetlands constructed primarily for water quality treatment include recreation, nature study, aesthetics, and education. An increasing number of treatment wetlands are being designed as attractive and informative park-like areas. Due to their urban setting, stormwater treatment wetlands such as Greenwood Park in Orlando, Florida, and Coyote Hills located east of San Francisco Bay, California, are heavily visited and utilized for field trips and other educational purposes. Two wetlands constructed for wastewater treatment (Arcata, California, and Ironbridge, Florida) are important recreational areas offering jogging and bird watching opportunities as major ancillary benefits. The human uses listed above (including the desirable benefit of just knowing that the wetland and its wildness are still there at the edge of town) are perhaps the most important factors in the popular support for protecting and enhancing the existing wetland resource base.

DESIGN FOR ANCILLARY BENEFITS

General wetland design features affecting the secondary functions of NPS wetlands were presented above. This section focuses on specific design considerations to provide those wetland features and their resulting functions, while simultaneously optimizing the primary functions of flood attenuation and pollutant removal.

Wetland design includes decisions concerning sitting, cell size and configuration, water flow and depth control, planting, and stocking with wildlife species. Because the primary function of the wetland *a priori* is NPS control, the water source is not a design decision; however, the level of pretreatment is an important item for consideration. Each of these topics is discussed briefly below.

Wetland Siting

NPS control wetlands can be sited close to individual stormwater sources or further downstream in a watershed, intercepting a tributary. One effect of wetland sitting on the resulting wetland functions is the quantity and timing of water in the system. Wetlands sited in headwater areas generally will receive more irregular and less dependable inflows, potentially resulting in prolonged dry conditions (unless soils are very impermeable or groundwater levels are normally high). This relative lack of flooding will reduce the quantitative magnitude of ancillary wetland values such as primary and secondary production. Maintenance of a healthy stand of wetland dependent vegetation may be difficult and upland or transitional species may eventually predominate. This type of system certainly will have some wetland values and may support differing faunal assemblages seasonally; however, the overall production of wetland-dependent species will likely be lower than in a perennially flooded wetland.

Sitting of the constructed wetland further downstream in the watershed may result in a different constraint, namely too much water during stormwater periods. Design of a downstream system may be "offline," allowing capture of only a portion of flood flows to prevent the washout of vegetation and berms. A series of offline constructed wetlands, each capturing a portion of the storm flows, can be used to deal with high storm volume. Wetlands located downstream in a watershed have a higher potential to have perennial water and higher ancillary food-chain benefits due to more constant base flows and generally higher groundwater levels.

Wetland sitting also may be very important because of other concerns such as the benefits of having adjacent donor wetlands for plant seeds and wildlife; adjacent undeveloped uplands to provide greater habitat diversity; or the importance of human contact and aesthetics. These sitting issues are dependent upon project-specific goals.

Cell Size and Configuration

Wetland cell size is based primarily on water quality treatment and cost considerations. Larger cells require less berm construction per unit area and fewer inlet/outlet structures, resulting in reduced project costs per area. For example, larger constructed wetlands (greater than 100 ha) may cost about $10,000/ha to construct, while smaller constructed wetlands may cost about $50,000/ha. Cell size may affect usage by some larger wildlife species but it has minimal effects on plant productivity or secondary production of most wetland animals (Sather and Smith, 1984). A greater berm to cell area ratio is typical of smaller wetland cells and may result in increased edge effect and increased nesting and feeding habitat for many mammal and bird species, as long as berms are infrequently mowed or visited.

Islands surrounded by marsh or open water provide excellent habitat for nesting waterfowl. Islands with trees are the preferred nesting habitat for wading bird rookeries in many wetlands. Nesting islands for waterfowl should be only about 0.6 m above normal high water, while higher and lower islands also may be valuable for other species for feeding, resting, or nesting.

Inclusion of open water areas not only improves the water quality treatment potential of constructed wetlands (Knight and Iverson, 1990), but also greatly enhances their ancillary benefits for wildlife. Mallard duck production is maximum in wetlands with approximately equal areas of marsh and open water (Ball and Nudds, 1989). Open water areas can be created by excavating a minimum of about 1.5 m below normal water level, and deeper excavations can provide greater hydraulic residence times and fisheries habitat. To prevent hydraulic shortcircuiting, open water areas should not be connected along the flow path, but rather interspersed with densely vegetated shallow marsh habitat (about 0.3 m average depth or less).

Cell number and configuration in series or in parallel is a major consideration for treatment capability and operational flexibility. These design considerations affect ancillary wetland benefits primarily through their importance for water flow and depth control.

Water Flow and Depth Control

Water depth and flow rate are important factors affecting dissolved oxygen in wetlands. Higher flow rates resulting from shallow water conditions tend to provide higher dissolved oxygen concentrations in marsh areas due to the increased influence of atmospheric recrearation. Higher dissolved oxygen levels generally, result in higher secondary production of aquatic invertebrates and vertebrates, increasing these ancillary wetland benefits. While deeper water in a marsh area may increase hydraulic residence time, this longer reaction time in many cases does not result in enhanced water quality, treatment (oxidation of organic matter and ammonia) because of the resulting reduction of dissolved oxygen.

Water depth is one of the main determinants affecting wetland plant growth. High water levels will stress growth of emergent macrophytes and encourage dominance by floating or submerged plants or algae. The hydrological tolerance range and optimum hydroperiod should be known for any desired vegetation type and closely adhered to in design of water-level control structures. Ideal design for water-level control allows water levels to be varied from zero (drained) to the maximum depth tolerance of desired wetland plant communities. Stop logs or weir plates should be of a type that effectively seal against leaks to help maintain water levels during periods of limited inflows. Multiple inlet and outlet weirs between adjacent cells allow greatest hydroperiod control flexibility.

Vegetation Planting

The selection of appropriate plant species for inclusion in a wetland constructed for NPS control will greatly influence ancillary benefits such as primary and secondary productivity. Improper plant species selection will result in low productivity, and a lengthy adaptive period may be necessary until available plant species, either planted or occurring naturally, rearrange themselves according to hydrologic factors. High plant diversity frequently can be achieved by natural colonization from existing soil seed banks in an area graded and shallowly flooded, or by using the technique of spreading muck and associated propagules from a donor wetland area (Gilbert et al., 1981). Wetland vegetation establishment is most rapid with closely spaced plants (less than 1 m on centers), planted during the growing season (Lewis and Bunce, 1980; and Broome, 1990). No marsh species is totally unutilized by wildlife, either directly for food or shelter or indirectly through the detritus food chain; therefore, expensive management to exclude "noxious species" or to select for favored species may lower overall wildlife utilization in favor of optimizing specific wildlife species. Burning marshes may be good management for waterfowl species but may be poor management for other bird species, fish, or small mammals.

Wildlife Stocking

Stocking constructed wetlands with mosquito fish (*Gambusia affinis*) has repeatedly been found to provide effective mosquito control as long as deeper water refuge areas, periodically free of floating vegetation, are available to provide perennial habitat. Mosquito fish, in turn, are an important forage fish for wildlife. Other forage fish that can be easily stocked (such as shiners, minnows, shad, and sunfish) contribute to a potentially long food chain of sport fish, reptiles, wading birds, waterfowl, and raptors.

Other food chain components may also be stocked or allowed to naturally immigrate to the constructed wetland. Significant fur bearer populations (otter, mink, muskrat, or nutria) can be supported in highly productive constructed wetlands. As with many wetland-dependent birds, a mixture of open water and marsh habitat is also essential for enhancement of these mammal species. Stocking of constructed wetlands located some distance from existing wetland habitat may be very important in quickly establishing ancillary benefits for wildlife and aesthetic uses.

Inflow Pretreatment

In conjunction with wetland hydroperiod, water quality is one of the key determinants of a wetland's form and function. Primary water quality characteristics affecting wetland plant communities are nutrients (especially nitrogen and phosphorus), suspended sediments, salts, pH, and temperature. Other than nutrient concentrations, these same water quality characteristics also greatly influence faunal populations. Inflow concentrations of biodegradable solids and the ammonia form of nitrogen may have indirect effects on wetland flora and fauna through their control of dissolved oxygen concentration in wetlands.

Within the normal range of fresh surface waters that might be captured by wetlands constructed for NPS control, only suspended solids, nutrients, and salts in a few areas are likely to occur at levels that might be detrimental to specific ancillary wetland goals. High suspended solids loads, if released within the constructed wetland, may smother plant growth in inflow areas (Kuenzler, 1990). This potential problem generally can be controlled by the use of a pretreatment grassed swale, a high-maintenance pretreatment wetland cell, or a pond prior to the habitat wetland (Livingston, 1989). If mineral suspended solids (clays, silt, and sand) are trapped in a pretreatment area, disruptive maintenance of the NPS wetland may be much less frequent.

Nutrient levels also will be reduced somewhat by passage though swales and ponds. Nutrient reduction may not be desired for a constructed wetland designed for the ancillary benefit of wildlife enhancement. As noted earlier, higher nutrients generally result in higher primary productivity of wetland plants and higher resulting wildlife utilization and production.

Human Access

Inclusion of boardwalks and blinds may greatly enhance the ancillary benefits of recreation and scientific research in a constructed wetland. While public access to a created wetland may disturb wildlife populations, these functions can be compatible if control of access to certain areas is maintained and islands for roosting and nesting are provided.

DESIGN TO MINIMIZE POTENTIAL PROBLEMS

Wetlands utilized for NPS control may create nuisance conditions that potentially negate their positive ancillary benefits. These nuisance conditions fall into two broad categories: (1) conditions that are a nuisance or hazardous to humans and (2) conditions that are hazardous to plants and wildlife. Each of these types of potential problems and their potential corrective measures are described below.

Nuisances to Society

Historically, wetlands were considered to be nuisance areas, harboring disease, poisonous reptiles, and noxious conditions. Wetlands drainage certainly saved many thousands of lives prior to the age of miracle drugs for prevention of malaria and yellow fever. The irradication of most insect-transmitted diseases as well as effective biting-insect control through biological and chemical agents has softened society's adaptive wetland loathing. However, periodic outbreaks of encephalitis continue to result in warnings from public health officials concerning creation of wetlands for water quality treatment (Wellings, 1986). Mosquito control districts continue to get dozens of calls following rainy spells that result in the synchronized emergence of biting adult mosquitoes. Golfers and homeowners continue to insist on mowed margins along stormwater ponds in the southern United States due to concerns about the occurrence of poisonous snakes.

Mosquito control using mosquito fish is relatively easy in constructed wetlands as long as perennial water areas exist and strongly anoxic conditions are avoided (Martin and Eldridge, 1989). Many wetlands receiving only NPS loadings may periodically go dry, resulting in total loss of mosquito fish populations. Without natural or intentional restocking with mosquito fish, these constructed wetlands are likely to result in significant nuisance conditions if located near populated areas.

Wetland-dependent venomous snakes such as the water moccasin (*Ankistrodon pisivorous*) and alligators (*Alligator mississippiensis*) are attracted to created wetlands in the southeastern United States that have high vertebrate and fish productivity. Warning signs, boardwalks, and mowed hiking trails are generally adequate to prevent the potential loss of recreational values due to poisonous snakes and other dangerous reptiles.

Because wetlands are infrequently used for water contact recreation, direct disease transmission by water-borne pathogens is unlikely (Shiaris,

1985). Wetlands used for NPS control are not generally utilized for human water supply. Due to potential toxic metals and pesticides sometimes found in NPS waters, potable use must be carefully monitored. One potential toxin pathway to humans is consumption of contaminated fish or wildlife from a treatment wetland. As discussed more fully in the next section, toxic metals and organic compounds must be prevented from accumulating to toxic concentrations in NPS wetlands to protect wildlife and humans. Forethought in wetland design and water pretreatment, as well as periodic monitoring of wetland water quality, is necessary to prevent any potential detrimental food chain effects in humans who consume wetland plants or animals.

Environmental Hazards

Environmental hazards that may occur when wetlands are used for NPS control include effects due to high loadings of pollutants that are normally subsidies (for example, too much water, organic matter, or nutrients) and effects resulting from metals, pesticides, and other potentially toxic chemicals.

Typical environmental changes resulting from too much water, biodegradable organic matter, or hypereutrophication due to high nutrient inputs are generally attributable to the direct or indirect action of all of these compounds to lower dissolved oxygen concentrations. Drastically lowered dissolved oxygen can result in significant losses of wetlands vegetation and fauna. Limiting hydrologic and oxygen-demanding loadings to wetlands is relatively easy to accomplish. While fairly exotic wildlife impacts such as avian cholera and botulism (Friend, 1985) and parasitic nematodes (Spalding, 1990) are possible due to depressed oxygen concentrations, these have not been shown to be a widespread threat to the use of wetlands for NPS control.

Other pollutants removed by NPS treatment wetlands are conservative from the standpoint of accumulation and storage in the wetland sediments, plants, and wildlife. Metals and organochlorine compounds are among the most likely conservative pollutants that invariably accumulate in wetlands receiving stormwater or agricultural drainage waters. While both direct and indirect environmental effects of toxins are possible in NPS wetlands, there is little evidence that these potential issues represent a real limitation on the use of wetlands for flood control and water quality management (Chan et al., 1982). However, where inflow concentrations of toxins are a concern, or in specific environments where indirect toxic or lethal conditions may develop, wetland planning and design must seek to minimize wildlife impacts.

Heavy metals generally follow one of two behaviors in wetlands (Gardner, 1980; Rudd 1987). Metals such as arsenic, cadmium, chromium, nickel, and zinc are quickly concentrated in soils and plants compared with water concentrations, primarily through direct adsorption and absorption. Rooted plants also acquire some metals via uptake from soils. Bioconcentration factors between the water and plant tissues are in the range from 100 to 1,000 times. In spite of these increased concentrations in plants, concentrations of these metals are not magnified through the food chain; dry weight concentrations decrease at higher food chain levels so that bioconcentration factors between water concentrations and fish tissues are less than 100 times (USEPA, 1986). These metals essentially reach saturation levels in tissue based on water concentrations, and additional uptake is matched by tissue metal losses, resulting in a relatively constant body burden. As long as source control or pretreatment prevents consistently high concentrations of these metals in the wetland influent, levels toxic to biota are unlikely to occur.

Microbially methylated forms of mercury and lead bioaccumulate in plants and also become concentrated through food-chain biomagnification. This concentration occurs because these metal-organic complexes have an affinity for lipids and are accumulated in tissues during the lifetime of an organism. Organochlorines such as DDT and dioxins biomagnify in the wetland food chain because of the same affinity for fats.

As with other metals, excretion and release mechanisms also exist for methylated mercury, lead, and organochlorines. Steady-state levels can be reached that do not result in toxic effects as long as input concentrations to the wetland are low. Safe input concentrations are not clearly known; therefore, compliance with published water quality standards is perhaps the best recommendation for these compounds. The existing water quality, criteria for metals are intended to protect the most sensitive organisms within the waters of the United States. In wetlands, these concentrations are protective of invertebrates or fish that may reside in the vicinity of the inflow. Since metal concentrations in NPS waters are frequently above these protective criteria, a tradeoff is necessary if wetlands are to be used for this purpose and the potential ancillary benefits are to be realized. Perhaps the most difficult issue to address is whether the creation of habitat for hundreds or thousands of ducks and wading birds is ample justification to exceed wetland surface water metal concentrations that are potentially chronically toxic to invertebrates or larval fish but that will not result in chronically toxic conditions for adult fish or birds. Development of biological criteria for wetlands to replace existing water quality criteria developed for streams and lakes may provide an answer to this regulatory dilemma (USEPA, 1990).

To date, there are no generally known incidences of conditions in treatment wetlands (municipal wastewater and stormwater) that have resulted in lethality to fish or other wildlife. The only documented cases of toxicity to wetlands wildlife known to this author are releases from hazardous waste sites (for example, USEPA, 1989), and discharges of agricultural irrigation return flows in the western United States (Willard and Willis, 1988; Deason, 1989). Research with agricultural drainage water at the Kesterson National Wildlife Refuge has recorded effects including vegetation changes, losses of species, fish die-offs, and acute and chronic effects on birds, primarily from highly concentrated levels of selenium. While sites such as Kesterson, Imperial Valley, Stillwater, and other wildlife refuges dependent on agricultural waters were not designed for treatment but rather for habitat, they serve as poignant examples of what must be avoided in the design of new wetland treatment systems.

SUMMARY AND RECOMMENDATIONS

The potentia] ancillary benefits derived from the wise use of wetlands for NPS control are great. While protecting the Nation's surface water resources, NPS treatment wetlands can provide a significant interest in the Nation's wetland resource base, provide a high level of additional food-chain support for wildlife and humans, maintain genetic diversity, provide export to neighboring ecosystems, and enhance society through provision of aesthetics, recreation, and education. Maximizing these potential benefits through the planning and design of new wetland treatment systems, or through modified operations and management of existing systems, is a worthy goal. Minimizing the potential negative effects of this emerging technology is an equally important requirement.

The state of our knowledge concerning wetland functions and what environmental factors control these functions is rapidly advancing due to wetlands research efforts worldwide. The practical application of this diffuse knowledge base is difficult because of the lack of an organized synthesis concerning wetlands management. The sole recommendation of this author is that the existing knowledge concerning wetlands management for ancillary benefits be organized into practical guidelines, to be periodically updated, for the planning, design, and operation of wetland treatment systems.

REFERENCES

Ball, J.P. and T.D. Nudds, 1989. Mallard habitat selection: An experiment and implications for management. pp. 659-671. In: R.R. Sharitz and J.W. Gibbons (eds.), Freshwater Wetlands and Wildlife, Proceedings of a Symposium Held March 24-27, 1986, Charleston, S.C. US Department of Energy, Oak Ridge, TN.

Broome, S.W., 1990. Creation and restoration of tidal wetlands of the southeastern United States. pp. 37-72. In: J.A. Kusler, and M.E. Kentula, Wetland Creation and Restoration. The Status of the Science. Island Press, Washington, DC.

Brown, S., M.M. Brinson, and A.E. Lugo, 1979. Strategies for protection and management of floodplain wetlands and other riparian ecosystems. pp. 17-31. In: R.R. Johnson and J.F. McCormick (eds.), Proceedings of the Symposium, Structure and Function of Riparian Wetlands, December 11-13, 1978, Callaway Gardens, GA. USDA, Washington, DC.

Chabreck, R.H, 1979. Wildlife harvest in wetlands of the United States. pp. 618-631. In: P.E. Greeson, J.R. Clark, J.E. Clark (eds.), Wetland Functions and Values: The State of Our Understanding, Proceedings of the National Symposium on wetlands, Lake Buena Vista, FL, November 7-10, 1978. American Water Resources Association, Bethesda, MD.

Chan, E., T.A. Bursztynsky, N. Hantzsche, and Y.J. Litwin, 1982. The Use of Wetlands for Water Pollution Control. EPA-600/600/2/12-82-086. 261 pp.

Deason, J.P., 1989. Impacts of irrigation drainwater on wetlands. pp. 127-138. In: D.W. Fisk (ed.), Proceedings of the Symposium on Wetlands: Concerns and Successes. September 17-22, 1989, Tampa, FL, American Water Resources Association, Bethesda, MD.

Erwin, K.L., 1990. Wetland evaluation for restoration and creation. pp. 429-458. In: J.A. Kusler and M.E. Kentula, Wetland Creation and Restoration: The Status of the Science. Island Press, Washington, DC.

Friend, M., 1985. Wildlife health implications of sewage disposal in wetlands. pp. 262-269. In: P.J. Godfrey, E.R. Kaynor, S. Pelczanski and J. Benforado (eds.), Ecological Considerations in Wetlands

Treatment of Municipal Wastewaters. Van Nostrand Reinhold Co., New York, NY.

Gardner, W.S., 1980. Salt marsh creation: Impact of heavy metals. pp. 126-131. In: J.C. Lewis and E.W. Bunce (eds.), Rehabilitation and Creation of Selected Coastal Habitats: Proceedings of a Workshop. U.S. Fish and Wildlife Service, FWS/OBS-80/27.

Gilbert, T., T. King, and B. Barnett, 1981. An Assessment of Wetland Habitat Establishment at a Central Florida Phosphate Mine Site. U.S. Fish and Wildlife Service, FWS/OBS-81/38. 95 pp.

Golet, F.C., 1979. Rating the wildlife value of northeastern freshwater wetlands. pp. 63-73. In: P.E. Greeson, J.R. Clark, and J.E. Clark (eds.), Wetland Functions and Values: The State of Our Understanding, Proceedings of the National Symposium on Wetlands, Lake Buena Vista, FL, November 7 - 10, 1978. American Water Resources Association, Bethesda

Gosselink, J.G. and R.E. Turner, 1978. The role of hydrology in freshwater wetland ecosystems. pp. 63-78. In: R.E. Good, D.F. Whigham, and R.L. Simpson (eds.), Freshwater Wetlands, Ecological Processes and Management Potential. Academic Press, New York, NY.

Greeson, P.E., J.R. Clark, and J.E. Clark (eds.), 1979. Wetland Functions and Values: The State of Our Understanding, Proceedings of the National Symposium on Wetlands, Lake Buena Vista, FL, November 7-10, 1978. American Water Resources Association, Bethesda, MD, 674 pp.

Guntenspergen, G.R., F. Stearns, and J.A. Kadlec, 1989. Wetland Vegetation. pp. 73-88. In: D.A. Hammer (ed.), Constructed Wetlands for Wastewater Treatment: Municipal, Industrial, and Agricultural. Lewis Publishers, Chelsea, MI.

Hair, J.D., G.T. Hepp, L.M. Luckett, K.P. Reese, and D.K. Woodward, 1978. Beaver pond ecosystems and their relationship to multi-use natural resource management. pp. 80-92. In: R.R. Johnson and J.F. McCormick (eds.), Strategies for Protection and Management of Floodplain Wetlands and Other Ripariarn Ecosystems, Proceedings of the Symposium, December 11-13, 1978, Callaway Gardens, GA. USDA, Washington, DC.

Hammer, D.A. (ed.), 1989. Constructed Wetlands for Wastewater Treatment: Municipal, Industrial, and Agricultural. Lewis Publishers, Chelsea, MI.

Knight, R.L., 1990. Wetland systems. In: S. Reed (ed.), Natural Systems for Wastewater Treatment, Water Pollution Control Federation, Alexandria, VA, MOPFD-16, 270 pp.

Knight, R.L. and M.E. Iverson, 1990. Design of the Fort Deposit, Alabama, constructed wetlands treatment system. pp. 521-524. In: P.F. Cooper and B.C. Findlater (eds.), Constructed Wetlands in Water Pollution Control. IAWPRC, Pergamon Press, Oxford, UK.

Kuenzler, E.J., 1990. Wetlands as sediment and nutrient traps for lakes. pp. 105-112. In: Proceedings of a National Conference on Enhancing the States' Lake and Wetland Management Programs, May 18-19, 1989, Chicago, IL.

Kusler, J.A. and M.E. Kentula, 1990. Wetland Creation and Restoration: The Status of the Science, Island Press, Washington, DC. 591 pp.

Kusler, J.A. and P. Riexinger (eds.), 1986. Proceedings of the National Wetland Assessment Symposium, Portland, ME, June 17-20, 1985. Association of State Wetland Managers, Chester, VT. 331 pp.

Lewis, J.C. and E.W. Bunce (eds.), 1980. Rehabilitation and Creation of Selected Coastal Habitats: Proceedings of a Workshop, U.S. Fish and Wildlife Service, FWS/OBS-80/27.

Livingston, E.H., 1989. Use of wetlands for urban stormwater management. pp. 253-262. In: D.A. Hammer (ed.), Constructed Wetlands for Wastewater Treatment Municipal, Industrial, and Agricultural. Lewis Publishers, Chelsea, MI.

Martin, C.V. and B.F. Eldridge, 1989. California's experience with mosquitoes in aquatic wastewater treatment systems. pp. 393-398. In: D.A. Hammer (ed.), Constructed Wetlands for Wastewater Treatment: Municipal, Industrial, and Agricultural. Lewis Publishers, Inc., Chelsea, MI.

Mitsch, W.J. and J.G, Gosselink, 1986. Wetlands. Van Nostrand Reinhold Co., New York, NY. 539 pp.

Nash, R., 1978. Who loves a swamp? pp. 149-156. In: R.R. Johnson and J.F. McCormick (eds.) Strategies for Protection and Management of Floodplain Wetlands and Other Riparian Ecosystems, Proceedings of the Symposium, December 11-13, 1978, Callaway Gardens, GA. USDA, Washington, DC.

Nixon, S.W. and V. Lee, 1986. Wetlands and Water Quality. A Regional Review of Recent Research in the U.S. on the Role of Freshwater and Saltwater Wetlands as Sources, Sinks, and transformers of Nitrogen, Phosphorus, and Various Heavy Metals, U.S. Army

Corps of Engineers, Wetlands Research Program, Technical Report Y-86-2.

Reimold, R.J. and M.A. Hardisky, 1979. Nonconsumptive use values of wetlands. In: P.E. Greeson, J.R. Clark, and J.E. Clark (eds.), Wetland Functions and Values: The State of Our Understanding, Proceedings of the National Symposium on Wetlands, Lake Buena Vista, FL, November 7-10, 1978. American Water Resources Association, Bethesda, MD.

Richardson, B. (ed.), 1981. Selected Proceedings of the Midwest Conference on Wetland Values and Management, St. Paul, MN, June 17-19, 1981, Freshwater Society, Navarre, MN. 660 pp.

Richardson, C.J., 1979. Primary productivity values in freshwater wetlands. pp. 131-145. In: P.E. Greeson, J.R. Clark, and J.E. Clark (eds.), Wetland Functions and Values: The State of Our Understanding, Proceedings of the National Symposium on Wetlands, Lake Buena Vista, FL, November 7-10, 1978. American Water Resources Association, Bethesda, MD.

Rudd, T., 1987. Scope of the problem. pp. 1-29. In: J.N. Lester (ed.), Heavy Metals in Wastewater and Sludge Treatment Processes. Volume 1. Sources, Analysis, and Legislation. CRC Press, Boca Raton, FL.

Sather, J.H. and R.D. Smith, 1984. An Overview of Major Wetland Functions and Values. U.S. Fish and Wildlife Service, FWS/OBS-84/18.

Shiaris, M.P., 1985. Public health implications of sewage applications on wetlands: microbiological aspects. pp. 243-261. In: P.J. Godfrey, E.R. Kaynor, S. Pelcrnnski, and J. Benforado (eds,), Ecological Considerations in Wetlands Treatment of Municipal Wastewaters. Van Nostrand Reinhold Co., New York, NY.

Smardon, R.C., 1988. Aesthetic, recreational, landscape values of urban wetlands. pp. 92-95. In: J.A. Kusler, S. Daly, and G. Brooks (eds.), Proceedings of the National Wetland Symposium on Urban Wetlands, June 26-29, 1988, Oakland, CA. Association of Wetland Managers, Berne, NY.

Smith, L.M., R.L. Pederson, and R.M. Kaminski, 1989. Habitat Management for Migrating and Wintering Waterfowl in North America. Texas Tech University Press, Lubbock, TX. 560 pp.

Spalding, M.G., 1990. Antemortem diagnosis of Eustrongylidosis in wading birds (Ciconiiformes). Colonial Waterbirds, 13: 75-77.

U.S. Environmental Protection Agency, 1986. Superfund Public Health Evaluation Manual. EPA 540/1-86/060. 145 pp.

U.S. Environmental Protection Agency, 1989. Water Quality and Toxic Assessment Study, Mangrove Preserve, Munisport Landfill Site, North Miami, Florida. Environmental Services Division, Athens, GA.

U.S. Environmental Protection Agency, 1990. Biological Criteria. National Program Guidance for Surface Waters. Criteria and Standards Division. EPA-440/5-90-004. 57 pp.

Weller, M.W., 1978. Management of freshwater marshes for wildlife. pp. 267-284. In: R.E. Good, D.F. Whigham, and R.L. Simpson (eds.), Freshwater Wetlands, Ecological Processes and Management Potential. Academic Press, New York, MY.

Weller, M.W., 1990. Waterfowl management techniques for wetland enhancement, restoration and creation useful in mitigation procedures. pp. 517-528. In: J.A. Kusler, and M.E. Kentula (eds.), Wetland Creation and Restoration: The Status of the Science. Island Press, Washington, DC.

Wellings, F.M., 1986. Letter to the Editor. Florida Water Resources Journal, 38: 38-39.

Wengrzynek, R.J. and C.R. Terrell, 1990. Using constructed wetlands to control agricultural nonpoint source pollution. In: Proceedings of the International Conference on the Use of Constructed Wetlands in Water Pollution Control, September 24-28, 1990, Churchill College, Cambridge, UK.

Willard, D.E.,and J.A. Willis, 1988. Lessons from Kesterson. pp. 116-121. In: J.A.Kusler,S. Daly, and G. Brooks (eds.), Proceedings of the National Wetland Symposium on Urban Wetlands, June 26-29, 1988, Oakland, CA. Association of Wetland Managers, Berne, NY.

CHAPTER 7

Regulations and Policies Relating to the use of Wetlands for Nonpoint Source Pollution Control

Sherri Fields, Wetlands Division, U.S. Environmental Protection Agency

ABSTRACT

The ability of wetlands to transform or trap nutrients and sediments has led to increasing attention to how wetlands can be used or their functions replicated to treat nonpoint sources of pollution. Protection, restoration, and creation of wetlands can be incorporated into nonpoint source pollution management strategies. There are, however, Federal regulations that prohibit the indiscriminate use of wetlands for water treatment. The Clean Water Act regulates all discharges into "waters of the United States" including wetlands. Restored wetlands are subject to the same protection and restrictions as natural wetlands. Created wetlands, on the other hand, are generally not considered "waters of the United States" if constructed solely for purposes of water treatment. Protection, restoration, and creation of wetlands provide opportunities to realize a number of functional benefits including water quality improvement.

INTRODUCTION

The many functions and values of wetlands are widely recognized, particularly with regard to their abilities to improve water quality. As a result, there has been increasing attention to how these natural systems can be used or their functions replicated to help treat nonpoint source (NPS) pollution. In considering this relatively new application of wetland functions, it is important to be aware of the current policies and regulations that affect the use of wetlands and wetland treatment systems. This paper addresses several topics. First, the relationship between wetlands and NPS control is discussed. Second, definitions of natural, restored, and created wetlands are provided. Third, the regulations and policies that govern the consideration of wetlands to treat NPS pollution are discussed. And fourth is some discussion of how protection and restoration can provide the incidental benefit of controlling NPS pollution.

WETLANDS AND NPS POLLUTION

In their natural orientation in the landscape, wetlands receive sediment and nutrients via runoff from uplands, and may retain or transform a portion of these inputs. Recognition of this capability by water quality engineers and others led to efforts to replicate these functions. Much of the impetus for using wetlands to treat NPS pollution stems from the successful use of wetland treatment systems to treat point source discharges, particularly wastewater effluent. Constructed wetlands have been used to treat primary or secondary treated wastewater effluent. Secondary wastewater effluent has been used to restore degraded wetlands. In addition, natural wetlands have been used to "polish" secondary treated wastewater effluent.

In the last couple of years, the Environmental Protection Agency (EPA) has recognized the important relationship between wetlands and NPS pollution control. In EPA's Nonpoint Sources Agenda for the Future (U.S. EPA, 1989), there is recognition of the ability of wetlands to help achieve NPS control. In 1990, EPA's Office of Wetlands Protection and Office of Water Regulations and Standards issued national guidance to encourage coordination between wetland and NPS programs (U.S. EPA, 1990a). In this guidance, EPA identified how State NPS control programs provide an opportunity to create, restore, and enhance wetland resources for water quality benefits. Although there is still a great deal to be learned about the relationship between wetland functions and NPS pollutants, the coordination of State and National programs can be a useful step toward the goal of maintaining the chemical, physical, and biological integrity of the Nation's waters.

DEFINITIONS

The use of wetlands for treatment of point source and NPS pollution has resulted in the introduction of a number of terms. Though the definitions of these terms have not been standardized across Federal agencies, EPA has attempted to be consistent in recent publications (i.e., National Guidance: Wetlands and Nonpoint Source (U.S. EPA, 1990a) and Water Quality Standards for wetlands: National Guidance (U.S. EPA, 1990b)).

"Natural" Wetlands are defined by EPA regulations as those areas that are inundated or saturated by surface or groundwater at a frequency and duration sufficient to support, and that under normal circumstances do support, a prevalence of vegetation typically adapted for life in saturated soil conditions (40 Code of Federal Regulations [CFR] Parts 122.2, 230.3, and 232.2). These systems are typically found as interfaces between open water and land or as isolated systems, such as prairie potholes. Wetlands exhibit a wide array of values and functions including wildlife habitat, flood water attenuation, nutrient retention or transformation, sediment retention, fish spawning habitat, and fish and shellfish habitat.

Restoration of Wetlands refers to returning a wetland from a disturbed or altered condition with lesser acreage or function, to a previous condition with greater acreage or function (U.S. EPA, 1990a). Restoration may involve reestablishing the original hydrology, vegetation, or other parameters that will restore the original appearance and/or functions of a wetland. Or restoration may involve removing a source of impact to a wetland. Related to restoration is enhancement, which refers to increasing one or more natural or created wetland functions. An example of enhancement is adding water to a drought impacted western wetland to increase its area and usefulness as habitat for migratory birds.

Creation of Wetlands refers to bringing a wetland into existence at a site where one did not formerly occur (U.S. EPA, 1990a). Wetlands are generally created from upland areas due to the intentional or nonintentional introduction of water or changes in surface elevation. Wetlands may be created intentionally, such as for waterfowl or wildlife habitat, as part of the Clean Water Act (CWA) Section 404 permit mitigation requirements, or incidentally, as a result of activities such as constructing highways.

Constructed Wetlands are a subset of created wetlands that are designed and developed specifically, for water treatment. Constructed wetland cells or systems are characterized by combinations of controlled water or effluent flow, specific vegetation (such as cattails or bulrushes), a waterproof liner, and other material resulting in the chemical improvement of the water flowing through. Constructed wetlands have been widely used as part of municipal wastewater treatment processes.

PRINCIPAL POLICIES

"Natural" Wetlands

Almost all natural wetlands are considered "waters of the United States" and, as such, are protected under several provisions of the CWA. Section 10(a) of the Act states that it is the objective of the CWA "to restore and maintain the chemical, physical, and biological integrity of the Nation's waters." Furthermore, it is stated in subsection (7) of this paragraph that it is the national policy that programs for the control of nonpoint sources of pollution be developed and implemented so that the goals of the Act can be met through the control of both point and NPS pollution. These sections provide the base justification for protecting natural wetlands from NPS pollution.

Section 319 of the CWA was enacted to address impacts to the Nation's waters from nonpoint sources. Under this funding incentive program, States are to develop NPS assessment reports and implement management programs. The purpose of the assessments is to identify "waters of the United States" that are impaired or threatened by NPS pollution, as well as the activities causing the impacts. Under this program, some States are developing and implementing enforceable provisions to address NPS pollution and to protect wetlands and other waters. In 1990, amendments to the Coastal Zone Management Act were passed that included requirements for development of enforceable management measures to address problems of NPS pollution in the coastal zone. These requirements will increase the enforcement leverage States need to address detrimental impacts.

Another important provision of the CWA that affects the use of natural wetlands for NPS control is Section 303. This Section requires States to adopt water quality standards that include designating uses for wetlands and other waters and to assign water quality criteria that will meet those uses. Although NPS discharges to wetlands are not directly regulated, States must prescribe use of best management practices (BMPs) to ensure compliance with the applicable State water quality standards. The refinement of BMPs is the mechanism in Section 303 to gradually reduce NPS impacts to all State waters.

Section 402 of the CWA has provisions for protecting "waters of the United States" from stormwater impacts. These impacts are typically a result of water quantity as well as water quality, as water is collected, transported, and discharged more efficiently through a stormwater system resulting in detrimental impacts to receiving waters, such as wetlands. The recent stormwater provisions regulate some of the formerly unregulated urban storm runoff, thus providing some regulation of impacts to wetlands and other waters.

Although nonpoint sources are not generally subject to Federal permits, under certain circumstances, Section 404 regulations may apply. Discharges of dredged or fill material into "waters of the United States" must be authorized by a permit under Section 404 of the CWA. For example, if a proposed NPS treatment activity involves discharging fill material into a wetland, a Section 404 permit may be necessary. Review of individual Section 404 permit applications includes review under the National Environmental Policy Act, and consideration of other applicable Federal laws and executive orders (such as the Endangered Species Act).

Restoration of Wetlands

Restoration of wetlands is addressed in several sections of the CWA. First, restoration of degraded wetlands is consistent with Section 101 of the CWA (see discussion of Section 101 above). Secondly, such wetlands are afforded protection for preservation of previous (before degradation) uses or functions under Section 303. Thirdly, if the source of degradation is a nonpoint source, the wetland should be identified in the State NPS assessment and included in the State's management program. And lastly, Section 404 regulations may apply if restoration efforts involve discharging fill material into a wetland.

Restoring degraded wetlands is a unique opportunity not only to improve the quality of our Nation's waters, but also to recapture other wetland functions that may have been lost. Restored wetlands provide some inherent water filtration functions that benefit adjacent waters. However, it should be remembered that these systems are considered to be "waters of the United States" and, therefore, NPS pollutants cannot be directed to them for treatment.

Creation of Wetlands

As defined above, created wetlands are a result of human activity, usually for a specific purpose. Wetlands can be created through diversions of water, discharges of treated effluent, or as a result of other activities, such as grading to lower soil surface elevation and allowing natural flows to inundate areas. Created wetlands often provide multiple benefits which may include water quality improvement. However, wetlands created for purposes other than wastewater treatment (e.g., mitigation of wetland losses under Section 404 or development of waterflow habitat) generally receive the same protections under the CWA as restored or natural wetlands. Any contributions of these created wetlands to NPS control cannot lead to degradation of the wetlands.

If wetlands are designed, constructed, and maintained for the sole purpose of water treatment, they are generally not considered "waters of the United States;" therefore, there are no applicable Federal regulations that govern their use (40 CFR, part 122.2). However, in these cases, water leaving the created wetland cannot significantly degrade or alter the water quality or other designated or existing uses of the adjacent waterbody.

Created wetlands can be used as part of a landscape-based approach to controlling NPS runoff. However, adequate consideration in design, maintenance, and monitoring must be given to prevent detrimental impacts to groundwater, birds, animals, and other biota.

HOW PROTECTION AND RESTORATION CAN PROVIDE INCIDENTAL NPS BENEFITS

NPS best management practices (BMPs) have for years been limited to structural measures that function as sediment traps or reduce erosion. More recently, nonstructural measures and processes are being developed and implemented under the auspices of pollution prevention. Pollution prevention is the use of processes, practices, or products that reduce or eliminate the generation of pollutants and wastes or that protect natural resources through conservation or more efficient utilization. Foremost, wetlands should be protected due to the many values and functions they provide. But, in addition, protection and restoration of wetlands are also acceptable management measures for preventing the impacts to water quality that result when wetlands are destroyed or degraded.

One principle of protection is avoiding impacts to wetland and riparian areas, when practicable, to maintain existing beneficial uses (functions) and to meet existing water quality standards. A similar principle applies to restoration: when conditions are appropriate, restoration of wetland and riparian areas is preferred over structural BMPs, or restoration can be used in conjunction with BMPs to gain not only water quality benefits but also additional benefits for "waters of the United States." The basic premise behind these approaches is that the benefit of improved water quality will be realized if wetlands and riparian areas are maintained (or restored) in the landscape to perform their natural functions. When this approach is used, additional BMPs, such as buffer zones, must be utilized to ensure that there is no adverse impact to wildlife using the wetlands and that the integrity of the wetlands will be maintained over time.

CONCLUSIONS

The ability of wetlands and riparian areas to filter and convert sediment and nutrients is wide!y accepted. Our increasing understanding of these wetland values has resulted in efforts to utilize or replicate natural systems. Wetlands, including degraded wetlands, are protected by provisions of the CWA from discharges of pollutants. On the other hand, the stipulations under Federal regulations are not as distinct for restored, created, or constructed wetlands used for NPS control.

It is important to note that the goals of NPS control programs and the goals of wetlands protection programs are not mutually exclusive. Protection and restoration of natural wetlands will result in the realization of NPS control benefits. Just as natural wetlands maintain water quality in a natural setting, created and constructed wetlands developed specifically for purposes of water treatment can be used to improve or maintain water quality of "waters of the United States."

The possibilities for helping restore nature's original capabilities or for replicating nature's processes are still evolving. Research on the functions and environmental effects of wetland treatment systems is critical to understanding the differences between natural and created systems, particularly with respect to how they can contribute to improvement of the physical, chemical, and biological integrity of our Nation's waters. Additional research is also needed to evaluate the potential for adverse impacts to wildlife and other biota utilizing wetlands receiving NPS pollution. Until we are able to know with more certainty the long term effects and other impacts of wetland treatment systems on the environment, we should proceed cautiously and refrain from the indiscriminate use of these systems for NPS pollution control.

REFERENCES

U.S. Environmental Protection Agency, 1987. Report on the Use of Wetlands for Municipal Wastewater Treatment and Disposal, Office of Water, Washington, DC.

U.S. Environmental Protection Agency, 1989. Nonpoint Sources Agenda for the Future, Office of Water, Washington, DC.

U.S. Environmental Protection Agency, 1990a. National Guidance: Wetlands and Nonpoint Source, Office of Water, Washington, DC.

U.S. Environmental Protection Agency, 1990b. Water Quality Standards for Wetlands: National Guidance, Office of Water, Washington, DC.

CHAPTER 8

The Role of Wetland Water Quality Standards in Nonpoint Source Pollution Strategies

Doreen M. Robb, U.S. Environmental Protection Agency, Office of Wetlands, Oceans, and Watersheds

ABSTRACT

States are required to develop water quality standards for their wetlands by the end of Fiscal Year 1993. Standards are vital to the protection of wetlands from a broad array of perturbations including nonpoint source (NPS) pollution. The natural water quality functions of wetlands make them potential components of NPS control strategies, but protection of wetland structure and functions takes precedence over their use in NPS control. Narrative biological criteria are one part of standards and can serve as a mechanism to address NPS pollution impacts. Criteria can also be used as a baseline to determine the effectiveness of best management practices. Numeric biocriteria are under development and will require additional research.

INTRODUCTION

How much of a sediment load can a wetland receive without being degraded? How much phosphorus can an inland marsh assimilate before it eutrophies or changes vegetation type? The answers to questions such as these are vital to States seeking to protect their wetlands from nonpoint source (NPS) pollution. Through the proper development and implementation of water quality standards (WQS) for wetlands, States can protect their wetlands and associated water quality functions. This paper will describe the components of WQS, then discuss the transition from narrative biological criteria to numeric biocriteria, and will finish by identifying U.S. Environmental Protection Agency (EPA) and State information needs for further development of WQS.

WATER QUALITY STANDARDS FOR WETLANDS

In July of 1990, EPA published national guidance requiring States to develop WQS for their wetlands by the end of Fiscal Year 1993 (USEPA, 1990). This will afford wetlands the same level of protection currently provided to other surface waters. Water quality standards consist of three parts; first, wetlands must have designated uses (e.g., aquatic life support, recreation, floodwater attenuation, groundwater recharge) consistent with the goals of the Clean Water Act. Next, narrative and numeric criteria are assigned to protect those uses. Narrative criteria are statements of attained or attainable conditions of a waterbody, and numeric criteria are numeric, usually chemical, endpoints that specify the maximum contaminant level that can be present without impairing the use of that waterbody. Third, each waterbody must have an antidegradation policy and implementation methods that protect previously existing uses of the waterbody as well as providing additional protection to higher quality and outstanding waters where the quality exceeds that necessary to maintain the uses.

Wetland standards are an important tool for States wishing to broaden the protection of their wetlands beyond minimizing the disposal of dredged and fill material as provided for by Section 404 of the Clean Water Act. Comprehensive narrative criteria, if implemented aggressively, can be used to protect wetlands from physical and hydrological modifications, including increased water flow, sedimentation, and nutrient overenrichment. By establishing criteria for a healthy wetland, a baseline exists against which changes in floral or faunal composition may be detected and evaluated. These baselines provide a basis for monitoring and assessing whether NPS pollution has detrimentally impacted a wetland. These changes can also be indications that best management practices (BMPs) are not achieving the desired result. BMPs can then be strengthened and refined until the waterbody in question is brought under compliance with water quality standards. The information needed to develop wetland standards and criteria will also be useful to designers of constructed wetland treatment systems in defining maximum loading rates for pollutants, and in monitoring system performance.

Water quality standards have evolved over the years. Originally, the 1965 Water Quality Act took a water quality-based approach by requiring States to develop WQS that specified levels of cleanliness for waters. Similar in concept to what we have today, numeric chemical criteria were developed to protect waters on a chemical-by-chemical basis. At the time, this water quality approach was not very effective since the necessary program infrastructure to enforce standards was lacking. Subsequently, the 1972 Clean Water Act established a technology-based approach that regulated individual point source discharges through National Pollutant Discharge Elimination System (NPDES) permits. These permits set guidelines for effluent limits and are basically "end-of-pipe" controls. Once the NPDES program was established and enforcement mechanisms were in place, the 1987 Clean Water Act (CWA) amendments re-established a water quality-based approach to supplement technology-based controls.

BIOLOGICAL CRITERIA

The 1987 Clean Water Act identified remaining serious pollution problems including toxic pollutants and NPS pollution. NPS impacts include sedimentation, eutrophication, hydrologic modification, bioaccumulation of toxics, increased turbidity and, subsequently, decreased light penetration. These impacts cannot be fully addressed on a chemical-by-chemical basis, and can lead to secondary impacts such as changes in vegetation type and associated biota. In an effort to address the primary impacts, EPA and the States have utilized narrative criteria. All States have adopted variations of aesthetic narrative criteria-the "free froms." "Free froms" are general statements, such as "free from debris, noxious odors, and taste." The development of narrative biological criteria (and numeric biocriteria in future years), however, is a new area of emphasis for EPA that will more effectively address secondary impacts.

Narrative Biocriteria

The development of narrative biological criteria and their application to NPS issues has important implications for wetlands. Narrative biological criteria are new requirements for all surface waters and are statements of attained or attainable condition necessary to protect the biological integrity of the waterbody. These criteria are flexible and can be written as very general or very specific statements. They can take the form of a "free from" statement, such as "free from activities that would substantially impair the biological community as it naturally occurs due to physical, chemical, and hydrologic changes." In their broadest sense, biological criteria protect the physical and structural components necessary for healthy aquatic habitat as well as the biota. For example, one State used more specific language to protect the natural hydrologic conditions of a wetland:

> "*Natural hydrological conditions* necessary to support the biological and physical characteristics naturally present in wetlands shall be protected to *prevent significant adverse impacts* on: (1) Water currents, erosion or sedimentation patterns; . . . (3) The chemical, nutrient and dissolved oxygen regime of the wetland; (4) The normal movement of aquatic fauna; . . . and (6) Normal water levels or elevations [emphasis added].

Narrative biocriteria can be effective in protecting wetlands from adverse impacts of NPS pollution if implemented effectively through BMPs. The examples described above are good first steps, and provide greater protection to wetlands than existed through numeric chemical criteria, but they have limitations and are difficult to enforce. A key point, however, is that the development of narrative biological criteria by States is based only on existing (and defensible) scientific information. For this reason, States should not have difficulty in developing narrative biocriteria immediately.

Numeric Biocriteria

The next step after the development of narrative biocriteria, however, is the development of numeric biocriteria. This is a future emphasis for EPA and the States and is based on the development of new scientific information. Numeric biocriteria have the potential to be more protective than narrative criteria because they are "hard" numbers and less subject to interpretation; therefore, they should be easier to enforce consistently. Currently, EPA is working on national guidance for rivers and streams; development of guidance for wetlands is slated for the future. An example of a numeric biocriterion for a coastal State is "vegetative diversity no greater than 2 species for salt marshes and no less than 25 species for a freshwater inland marsh" or "percent vegetative species change shall be no greater than 10 percent." Similar numbers could be derived for other components of the biota such as benthic invertebrates, breeding birds, and amphibians.

NPS CONTROLS AND WETLAND WATER QUALITY STANDARDS

Wetlands have an important role in the landscape through their ability to improve water quality by filtering, transforming, and accumulating pollutants and thereby protecting adjacent rivers, lakes, and streams. This "buffering" function, however, also encourages overuse, and this overuse can compromise these and other wetland functions, such as wildlife habitat and aesthetic and recreational values. While wetlands may be useful components of NPS pollution control strategies, the first goal must be protection of wetlands from pollution. EPA does not allow surface waters to be used as disposal sites for wastewater, and State water quality standards exist to ensure the protection of State waters, including wetlands. Consider the following examples; a State restores a degraded wetland for the purpose of slowing water that will flow off a new parking lot or highway, or a private landowner restores a degraded riparian area for the purpose of filtering sediment and nitrogen-enriched water from a nearby feedlot. At first glance, the "use" of a restored wetland in both of these examples protects the water quality of a nearby waterbody such as a lake or stream and that waterbody meets State water quality standards. However, in the case of the highway, toxics accumulate in the wetland in amounts that exceed toxics criteria, and in the feedlot example, the riparian area retains sediment that eventually modifies the flow of water through that area, changing vegetation and runoff patterns. In both of these examples, although the action benefitted the adjacent waterbody, it did so at the expense of the wetland and, therefore, those actions violated the water quality standards. Additional management practices may need to be put in place, such as vegetated grass filter strips to buffer the wetlands. Regardless of the solution used, the integrity of both the wetland and the adjacent waterbody must be protected.

MONITORING AND ASSESSMENT

State standards, however, are only one mechanism to control and minimize the degradation of wetlands by NPS pollution. The ability to detect impacts through biological monitoring is another important tool to prevent degradation and is critical to the effective use of water quality standards. Biological monitoring and assessment enables States to compile baseline information on wetland condition. This information can then be used in developing biological criteria. Once a State has established biological criteria for its wetlands, the State then has a regulatory mechanism to deal with impacts that violate State water quality standards.

Before States can establish a monitoring program for their wetlands, however, they need to know what to measure (i.e., indicators) and how to recognize an impact. For example, if a State knows to sample a particular benthic invertebrate, they also need to know what should be the expected population dynamics of that invertebrate in a particular type of wetland under "natural" and "perturbed" conditions. They need to know whether a change in vegetation is a natural community succession or an indicator of increased phosphorus loading causing an extremely diverse plant community to shift to a monotypic community. Such information will enable States to establish a wetlands monitoring program as well as aid them in future numeric biocriteria development.

FUTURE RESEARCH NEEDS

As States work to establish narrative biological criteria, EPA will be developing guidance for developing numeric biological criteria. Increasing the technical science base is necessary before numeric biocriteria can be developed. Examples of important research questions include:

- How do altered hydrology and sedimentation patterns impact wetlands and how does the biological community react to these changes?
- What should States measure to discern these changes?
- How much change is too much?

Such information is needed for all wetland types and regions so States can monitor for and recognize these impacts when they occur.

CONCLUSIONS

Wetlands have an important function in landscape water quality. As a result, they are often included in strategies for controlling NPS pollution. Water quality standards, however, apply to wetlands as well as to other waterbodies. Therefore, wetlands must be protected from NPS pollution through, for example, the use of BMPs such as upland buffers. State development of effective water quality standards for wetlands requires further research on indicators of wetland health, impacts and indicators of physical and hydrological alterations, and thresholds for sediment, nutrients, and toxics loading. These types of information will enable States to protect their wetlands through technically defensible water quality standards.

REFERENCE

U.S. Environmental Protection Agency, 1990. Water Quality Standards for Wetlands: National Guidance. EPA 440/S-90-011, Office of Water Regulations and Standards, USEPA, Washington, DC.

CHAPTER 9

Recommendations for Research to Develop Guidelines for the Use of Wetlands to Control Rural Nonpoint Source Pollution

Arnold G. van der Valk, Department of Botany, Iowa State University.
Robert W. Jolly, Department of Economics, Iowa State University.

ABSTRACT

Natural wetlands should not be used to reduce rural nonpoint source (NPS) problems. Properly designed restored or created wetlands, however, can be used for this purpose in many agricultural landscapes. Agricultural landscapes in which wetlands can be easily restored are the most suitable areas. Major technical issues that need to be resolved before effective and realistic guidelines can be developed for using restored wetlands to reduce NPS pollution include (1) the effects of contaminants, particularly sediments and pesticides, on restored wetlands; (2) the fate of organic contaminants in restored wetlands; (3) the development of site selection criteria; and (4) the development of design criteria. There also are many social, economic and political barriers to using restored wetlands. Social and economic issues that need to be resolved include (1) what is the most appropriate landscape unit for wetland restoration programs? (2) where should wetlands be sited? (3) who will make siting decisions? (4) how can landowner cooperation for restoration programs be obtained? (5) who will pay for wetland restorations? and (6) how cost effective is this approach? Watersheds are recommended as the natural landscape unit for planning, implementing, and administering restoration projects. Eight different research projects are identified: five technical projects (watershed-level demonstration projects, effects of contaminants on wetlands, sustainable loading rates for contaminants, landscape or watershed simulation models, and site selection and design criteria for restored wetlands) and three economic and social projects (attitudes of farmers and rural leaders, legal and public policy implications, economic costs and benefits).

INTRODUCTION

Rural nonpoint source (NPS) pollution is a landscape-level problem. Ultimately, eliminating, minimizing, or redirecting the movement of materials within agricultural landscapes is the only effective means of reducing NPS pollution to acceptable levels. Such a landscape-level solution should include a combination of in-field and off-field approaches. In-field approaches should include reduced inputs of nutrients and pesticides, best management practices to reduce soil erosion, and improved cropping systems. Off-field practices should include more appropriate land use, establishment of vegetated buffers or filter strips between farm fields and aquatic systems, and the creation of sinks for NPS contaminants near their points of origin. Off-field modifications, such as restoring or creating wetlands that act as nutrient sinks, offer potential as part of a comprehensive landscape-level strategy for NPS pollution reduction.

Wetlands may be the most cost-effective sinks for contaminants in many agricultural landscapes. Other papers in this volume review available data on wetlands as nutrient traps (Mitsch, Rodgers) and the construction of wetlands for wastewater treatment (Hammer). We will take it as a given that properly designed and constructed wetlands will intercept, store, and/or break down the various contaminants normally found in agricultural runoff (sediments, nutrients, pesticides). We do not wish to imply, however, that the fate of all contaminants in wetlands is fully understood or that the impact of most contaminants on wetlands is understood at all. In fact, very little about the latter is known. We also will assume that wetlands can be restored and created in appropriate places in the landscape as needed. In recent years, there has been enough work done on wetland restoration and creation that these per se are not issues. Much of the available information on wetland restoration and construction is summarized in Kusler and Kentula (1990) and Hammer (1989). Ancillary ecological and wildlife benefits from restoring wetlands in agricultural landscapes are reviewed by Knight (this volume). Although these ancillary benefits are important, and the creation of wildlife habitat has been the primary motivation for most wetland restorations so far, we will ignore them.

We do not recommend the use of natural wetlands as sinks for NPS contaminants. Natural wetlands in agricultural landscapes are usually rare and often are already at risk because their hydrology has been altered by regional drainage and because they may already receive significant inputs of agricultural runoff (Davis et al., 1981; Stuber, 1988; Neely and Baker, 1989). These few remaining wetlands are usually important habitat for many plant and animal species and are also important recreational areas, particularly for waterfowl hunters. Anyhow, the size and location of natural wetlands may not make them effective as sinks. If rivers, lakes, and reservoirs are to be protected from degradation by contaminants in agricultural runoff, natural wetlands should receive comparable protection. Wetlands have no magical properties that make them immune from degradation. Guidelines are currently being developed for water quality standards for wetlands (USEPA, 1990), and these should be applied to natural wetlands in agricultural landscapes. Restoring or creating wetlands as sinks in agricultural landscapes, however, makes sense, and all our comments assume that it is restored and created wetlands that will be developed as sinks, not natural wetlands.

We are defining restored wetlands as wetlands established in natural basins whose natural wetlands had been artificially drained. Restoring a wetland is usually done by blocking or removing the basin's man-made drainage system. A created wetland is a wetland established in an area that historically was not a wetland. Creating a wetland is done by excavating a suitable basin or constructing dikes, and it often requires redirecting water to the new basin. Creating a wetland is usually more expensive than restoring one. Constructed wetlands are a subset of created wetlands that were established specifically for wastewater treatment. To date most constructed wetlands have been built to treat point sources of human or animal waste or to treat urban storm runoff (Hammer, 1989). Although both restored and created wetlands should be equally effective for reducing rural NPS problems, we believe that the additional cost of created wetlands will make them, under many circumstances, uneconomical. So, we recommend that, at least initially, research focus on the use of restored wetlands in agricultural landscapes where wetland restorations can be done easily and cheaply.

Wetland restoration is not appropriate for all kinds of agricultural landscapes. Regions whose agricultural landscapes contain drained wetlands are the best candidates for this approach. As a first cut, suitable regions can be identified by inspection of Figure 1 , which is taken from Pavelis (1987) and reproduced in Dahl (1990). Figure 1 is a map of the United States that summarizes data on wetland drainage. As a second cut, Figure 1 can be superimposed on the surface water quality maps of the United States in Omernik (1977) or on fertilizer and atrazine application maps in Moody (1990). This will delimit regions that have lost most of their wetlands and that have the most polluted surface water. These areas are frequently congruent. The restoration of wetlands in these regions should result in significant improvements in water quality. Based on both wetland drainage and surface water quality criteria, the most suitable regions are the Midwestern corn belt (Iowa, Illinois, Indiana, and western Ohio), the lower Mississippi River valley (parts of Arkansas, Tennessee, Mississippi, and Louisiana), and southern Florida. There are three reasons why these areas are suitable: (1) they have a climate and topography suitable for wetlands; (2) land use in these regions is predominantly row-crop agriculture, the major source of NPS pollution; and (3) surface or subsurface drainage networks connecting drained wetlands can be used to channel runoff into restored wetlands. In other words, it is in these drained landscapes that restoring wetlands will be easiest and will do the most to improve water quality.

To determine what research needs to be done to ensure that effective guidelines are developed for using wetlands as sinks, we will first consider what are the major unresolved technical questions and then what are the major social and economic barriers to the implementation of this approach. Finally, we will recommend a number of research projects that we believe need to be completed before final guidelines for the implementation of this approach to NPS pollution abatement can be developed.

TECHNICAL ISSUES

Four general technical questions need to be resolved before the approach of restoring wetlands in watersheds to reduce rural NPS pollution can be implemented: (1) what effects will agricultural contaminants have on restored wetlands? (2) what is the fate of contaminants in restored wetlands? (3) where should restored wetlands be sited? and (4) what are appropriate design criteria for restored wetlands?

What are the Effects of Agricultural Contaminants on Wetlands?

There will be many constraints on the restoration of wetlands in agricultural landscapes that will largely decide how these restorations will be done. The most significant of these is cost. Most wetlands will need to be constructed at the least cost per wetland, farm, watershed, and region. There also will be strong economic and social pressure to take the minimum amount of land out of production. Consequently, most restored wetlands will have no, or only minimal, water control structures; will have little, if any, basin excavation; will not normally be planted or seeded to reestablish the vegetation; and will tend to be small. Many of these wetlands may be little more than small, shallow depressions that hold water seasonally. Initially, their vegetation composition, primary production, secondary production, and nutrient cycles will not resemble those of natural wetlands.

Because the primary source of water for restored wetlands will be agricultural runoff, often containing high amounts of nutrients, sediments, and pesticides, these wetlands may never become similar in composition, structure, or function to natural wetlands, which developed without such inputs. The impact of contaminants on restored and natural wetlands has been little studied (see Stuber, 1988), but it is not inconsequential. Reports of changes in the composition of natural wetlands due to increased inputs of nutrients from surrounding agricultural systems (e.g., the ongoing invasion of cattails into the northern Everglades in Florida) or inputs of pesticides on invertebrates (e.g., in prairie potholes; see Grue et al., 1986, 1988) suggest that there will be impacts. There are two recent compilations of information on the impacts of contaminants in agricultural runoff on wetlands, both emphasizing impacts on waterfowl (Sheehan et al., 1987; Facemire, no date). The information reviewed in these two publications shows unequivocally that contaminants impact both the flora and fauna of wetlands. What is not known is the extent to which contaminants, particularly pesticides and sediments, will alter the development of restored wetlands. Ongoing studies of the effect of sediment on litter decomposition and seed bank recruitment (van der Valk and Jurik, unpublished) indicate that contaminant effects deserve careful and detailed study. How closely restored wetlands will come to resemble natural wetlands is unknown, but it is unlikely that they will ever very closely resemble pristine natural wetlands.

What is the Fate of Agricultural Contaminants in Wetlands?

Agricultural runoff contains a complex and highly variable mix of dissolved and suspended contaminants (Neely and Baker, 1989). Its composition is a function of precipitation, topography, regional land use patterns, soil characteristics, fertilizer and pesticide application rates, tillage practices, etc. Of the three major classes of contaminants (nutrients, pesticides, and sediments) usually found in agricultural runoff, only the fate of nutrients in wetlands has ever been adequately studied (Howard-Williams, 1985; Bowden, 1987; Neely and Baker, 1989). The results of these studies indicate that denitrification is the major mechanism for the removal of nitrogen and that sedimentation is ultimately the major mechanism for the removal of phosphorus.

Restored wetlands, as do natural wetlands (Martin and Hartman, 1987; Phillips, 1989), will act as settling basins for sediment in agricultural runoff. What is a sustainable loading of sediment? What criteria should be used to decide what is a sustainable loading? Because very little research on sediment impacts on freshwater wetlands has been done (van der Valk et al., 1983), it is impossible to answer these questions.

There also is little known about the fate of most pesticides in any kind of wetland. Consequently, there is no way to determine what sustainable loading rates of different pesticides should be for restored wetlands. The effects of nutrient levels and sediment loads on pesticide degradation rates in restored wetlands deserves particular attention.

In short, the lack of information on the fate and sustainable loading rates of sediments, pesticides and, to a much lesser extent, nutrients in restored wetlands makes it impossible to develop guidelines for the use of these wetlands to control rural NPS pollution.

Where Should Restored Wetlands be Sited?

Figures 2 and 3 illustrate the two basic scenarios for the placement of wetlands in an agricultural landscape. In Figure 2, the wetland is placed at the base of the watershed, and in this position all water leaving the watershed will pass through it. The major advantage of this placement is that only one wetland must be established per watershed. In contrast, Figure 3 shows the wetlands distributed around the watershed so that each subwatershed has its own wetland. Individual wetlands would be much smaller and possibly easier to establish. The advantage of the distributed-siting pattern is that less runoff and erosion might occur in the whole watershed as a result of storing water and sediments high in the watershed (Novotny and Chesters, 1989). This could reduce the total area of wetland needed in a watershed.

Because so little work on this topic has been done, the optimal placing of wetlands in agricultural watersheds needs to be thoroughly examined. Optimal placement likely will vary from region to region because of different topographies, land use patterns, and layouts of surface and subsurface drainage networks.

What are Appropriate Design Criteria?

Ideally the use of wetlands in a watershed to improve water quality will be part of a more comprehensive plan to reduce NPS problems, including other off-field and improved in-field practices. Improved in-field practices, e.g., lower fertilizer application rates, could significantly reduce the area of wetlands needed in a watershed. Consequently, it makes sense to restore wetlands as part of a comprehensive watershed plan so that the area of wetland needed can be more realistically assessed.

In a water quality context, the single most important feature of a restored wetland is its size. The appropriate size of a restored wetland will depend on (1) the contaminant of greatest local concern that requires the longest residence time for its degradation, and (2) the percent reduction of this contaminant that is required seasonally, annually, or interannually. In operational terms, the size of a wetland should be determined by the expected total mass of various contaminants in the runoff entering it during some period and the tolerable or sustainable loading of these various contaminants per unit area of wetland. Both contaminant delivery and sustainable loading rates are difficult to quantify, the former because precipitation events are highly variable seasonally and interannually, and the latter because the sustainable loading for each contaminant is likely to be different. Turnover time, in turn, is a function of both precipitation patterns, wetland size, location of inflows and outflows, flow patterns within the wetland, etc. As noted previously, little is known about sustainable loading rates for restored wetlands for most agricultural contaminants.

Although many models have been developed to estimate the delivery of sediments and contaminants from nonpoint sources, they generally decrease in accuracy with increasing watershed size (Novotny and Chesters, 1989). The most reliable method for calculating deliveries of dissolved and sediment-bound contaminants to wetlands still needs to be determined. Eventually, from available models and more detailed delivery studies, some simple rule of thumb will need to be developed (e.g., one hectare of wetland is needed for each 100 hectares of watershed) or a simple delivery model will have to be chosen that can be used on a daily basis in the field.

One important design feature of wetlands that needs particular attention is maximizing residence times of runoff in wetlands. Currently, most wetland restorations are done to create waterfowl habitat. In landscapes where wetlands were drained with drainage tile networks, as in parts of the Midwest, wetlands are often created by interrupting drainage tiles. The standard way this is done is illustrated in Figure 4. Because inputs and outputs are physically adjacent, this design, although suitable for creating waterfowl habitat, is inappropriate for water quality purposes because residence time of water is effective

SOCIAL AND ECONOMIC ISSUES

We believe that social and economic considerations ultimately will decide whether landscape-level approaches, such as the restoration of wetlands in watersheds to reduce rural NPS problems, can and will be implemented. We have reached this conclusion because of our experiences with several agricultural landscape reconfiguration projects. Technical issues, although they are far from insignificant, are not nearly as daunting as the organizational, social, political, and economic issues that quickly arise when such projects are attempted. Agricultural landscapes will need to be managed to meet not only environmental goals but also economic and social goals. The cooperation of farmers and local community leaders is essential for the success of any landscape management program. Existing institutions and organizations at the county, State, and Federal levels will have a significant influence on the adoption and implementation of landscape management programs. Consequently, economic and social analyses are essential for understanding local attitudes, institutions and organizations, and economic constraints that must be addressed to ensure the success of this approach.

Among the major social and economic issues that need to be addressed are: (1) what is the most appropriate landscape unit for implementation of a wetland restoration program? (2) where should wetlands be sited? (3) who will decide where wetlands will be sited? (4) how can landowners' cooperation be obtained? (5) who will pay for the restoration and creation of these wetlands? and (6) how cost effective is this approach, from both a private and public perspective? The answers to these questions will undoubtedly require changes or adjustments in a variety of public policies and regulations. Relevant public policy and regulatory issues are discussed by Fields (this volume). Here, we will only mention a few economic and social issues that we feel need to be raised, researched, and debated prior to the implementation of any largescale, landscape-level program to restore wetlands.

What is the Most Appropriate Landscape Unit?

What the watershed is the natural unit for dealing with water-quality issues has long been recognized by researchers, many policy makers, and many administrators, but rarely by landowners and even more rarely by public officials. Each watershed is a unique natural geomorphological unit in which uplands and lowlands are linked hydrologically. The impact of land use and land-use changes on water quality can best be examined within a watershed, because each watershed is an isolated entity. In other words, human impacts on the environment can be most easily quantified within watersheds, and landscape modifications, such as restoring wetlands to improve water quality, can best be evaluated within watersheds. Economic externalities that arise from altering land use and management on individual farms can be more easily internalized if the basic unit is the watershed rather than the farm. It is not possible, however, to internalize all environmental costs at the watershed level. On the whole, the watershed is the most nearly ideal unit for integrating environmental, agricultural, economic development, and social programs in rural areas.

Can landscape-level programs be successfully organized on a watershed basis? To answer this question, other questions must be considered: What is the best way to do this? What kinds of economic and social incentives will be needed to induce farmers within a watershed to make collective decisions? Who will write guidelines for developing wetland restoration programs and who will approve programs for each watershed? How much of a problem will absentee landlords be?

Because property administrative, and political boundaries normally do not coincide with watershed boundaries, landowners and public officials normally do not think in terms of watersheds, and implementing soil erosion and water quality programs in a watershed framework remains politically and socially challenging. For example, the 1990 Food, Agriculture, Conservation and Trade Act retains, in effect, individual farms as the basic landscape unit. There are, however, already law's and programs that use watersheds as a landscape unit, e.g., the Watershed Protection and Flood Prevention Act, PL 566. PL 566 can provide

Federal funding to local sponsorship groups for flood control, water management and supply, grade stabilization and soil erosion prevention, and water quality projects. Determining how watersheds can be incorporated into public laws and regulations dealing with water quality issues requires careful consideration and investigation of existing and potential legal, economic, and regulatory options.

As noted, the landscape unit in agricultural programs, including most soil conservation programs, has been the farm or the individual tract. Because the placement of wetlands in a watershed is crucial if wetlands are to be effective as sinks for contaminants, it will be necessary to establish them in specific places. Historically, most soil conservation programs have been voluntary, i.e., there has been little targeting of resources to treat areas with the most severe erosion problems. Economic analyses of past programs suggest that this lack of targeting of resources has resulted in soil erosion programs that often were not cost effective in many regions of the country, including the Midwestern corn belt (Ribaudo et al., 1989). If the use of wetlands for water-quality programs is going to be cost effective, it is essential that it be done as part of a targeted, landscape-level, and comprehensive program.

Watersheds come in a variety of sizes. A determination must be made as to the appropriate size or stream-order for watersheds to be used for controlling NPS pollution. If too large a watershed is used, planning and implementation will become difficult because too many political and administrative units will be involved. If too small, the bureaucracy required to develop and implement a program may be too costly.

Where Should Wetlands be Located in Watersheds?

Where to restore wetlands in a watershed is, in part, a technical issue as outlined previously, but good is also a social, economic, and political issue. Figures 5 and 6 illustrate some advantages and disadvantages of siting wetlands at the base of a watershed or distributing them all over the watershed when farm boundaries are considered. Siting one large wetland at the terminus of a watershed (Figure 5) will involve dealing with fewer landowners, possibly only one. Other landowners in the watershed may not need to participate in the wetland restoration program at all. Basal siting means taking more land out of production on farms at the lower end of a watershed. Distributing the wetlands around the watershed (Figure 6) will mean having to deal with more landowners, but it means that less land may be taken out of production on any one farm. If the distributed approach is to work, nearly all landowners in a watershed will have to participate. The distributed model has the advantages of fairness, as no one landowner normally is doing more than what is required to treat the runoff from his or her land. Further, complex substitution procedures for landowners within a watershed may be avoided.

Who Will Make Siting Decisions?

Ideally, decisions about how to reduce agricultural NPS pollution in a watershed will be made voluntarily and collectively by the farmers who own property in the watershed and by appropriate local officials using technical alternatives provided by the Soil Conservation Service and Cooperative Extension Service, Public funds will undoubtedly continue to be used to install a variety of best management practices to reduce agricultural NPS problems. Restored and created wetlands are only one off-site control measure that could be incorporated into a watershed-level NPS pollution reduction plan.

How best to organize watershed-level programs that include wetland restorations is a key issue. One possibility would be for landowners within a watershed to organize a sponsorship group to develop restoration plans and request funding for them. This is an unlikely scenario in the immediate future because farmers know little about wetlands, their water-quality benefits, or their restoration. Information, education, and demonstration programs on the role of wetlands in agricultural watersheds are therefore needed. Alternatively, existing infrastructures at the county level, e.g., Soil and Water Conservation Districts, could be used to set up watershed-level committees within each county that would develop a plan for its watersheds. Since watersheds will often cross county lines, protocols for dealing with multicounty watersheds may need to be developed. Existing procedures and protocols for dealing with flood control and other multijurisdictional issues, however, may be adequate.

Although we think that it is preferable that both organizational planning and implementation be done at the watershed level, this is not essential. Wetland restoration programs must be developed at the watershed level, but the administration of these programs can be done using different existing administrative or political boundaries, e.g., counties. If administrative boundaries other than the watershed are used, coordination among different units within a single watershed will be essential. Although it will be more complex to use existing administrative boundaries to plan and implement watershed-level programs, the advantages of using the watershed as a unit will outweigh the added administrative burden that this causes.

What is the Best Way to Obtain Landowner Cooperation?

Getting farmers to think in terms of watersheds will require educating them about wetlands and water-quality issues and the technical reasons for organizing programs at the landscape level. Surveys of farmers' attitudes to watershed-level programs and wetland restorations are needed to determine what are the social and economic obstacles to such an approach and what kinds of incentives will be needed to get farmers to adopt them. These surveys will need to be done in several regions of the country; data should be collected and digested before any adjustments and changes in regulations and public policy are implemented. The insights gained from these surveys also should be used to develop educational programs for farmers on why wetlands are beneficial in agricultural landscapes and why watersheds are environmentally, socially, and economically the most natural and logical unit for dealing with NPS pollution problems.

Who will pay?

We already know the answer to this question, at least in general terms. Public funds primarily will compensate farmers for land taken out of production and restored to wetlands. At the Federal level, there are several existing programs within the USDA, including the new Wetlands Reserve Program, and within the Fish and Wildlife Service, including the various joint ventures that are part of the North American Waterfowl Management Plan, for wetland acquisition, restoration, and creation. Also, there are many State-level wetland protection, acquisition, and restoration programs as well as water quality protection programs that are potential sources of funding, e.g., the Reinvest in Minnesota (RIM) program. Lastly, there are many conservation organizations that will fund habitat restorations projects, e.g., Ducks Unlimited, Inc. How best to coordinate funding within Federal agencies and between the Federal government and State governments needs to be explored. How best to fund watershed restoration programs needs to be examined, and determination made as to what extent it is possible to internalize the costs of wetland restorations.

How Cost Effective is This Approach?

An evaluation of the potential cost effectiveness of this approach should be made before it is proposed on a national scale. Before this approach can be evaluated, however, water-quality goals for agricultural runoff need to be established nationally. Once national goals are proposed or established, spatial models of selected agricultural watersheds developed using a Geographic Information System (GIS) or other suitable modeling approach, can be used to determine the contribution of wetlands of various sizes and at various locations to achieving these goals in a particular watershed. NPS contaminant delivery models should be used with these spatial models initially to determine the location and size of the wetlands needed. Economic models applied to these spatial models can be used to determine the costs of restoring wetlands and the benefits derived from improved water quality. Although they all have shortcomings, models currently exist that can assess pollutant delivery to a wetland (e.g., Knisel, 1980; Young et al., 1987, Braden et al., 1989; Lane and Nearing, 1989; Novotny and Chesters, 1989). Linking watershed process and economic models, however, remains a major challenge.

PROPOSED RESEARCH

A great deal is known about the fate of nutrients in agricultural runoff in wetlands (Neely and Baker, 1989) and how to restore (Kusler and Kentula, 1990) and construct wetlands (Hammer, 1989). Nevertheless, as outlined above, using wetlands as contaminant sinks in agricultural landscapes raises many technical, economic, social, and legal questions for which there are no or only, partial answers. Among the most important topics that need to be studied before guidelines can realistically, be developed for implementing this approach are the fate and effect of contaminants other than nutrients, criteria for siting and designing restored wetlands, the rural population's attitudes toward landscape-level approaches to NPS reduction, and the best way to organize and fund wetland restoration programs at the watershed level.

The proposed research projects outlined below should be integrated as much as feasible. This can best be done by establishing several major regional studies to research relevant technical, social, and economic problems. Regional differences in landscapes (topography, soil characteristics, precipitation patterns, etc.), farming practices, attitudes toward wetlands, and economic realities may require different regional strategies.

There are eight topics that we think should be investigated. Each of these is described briefly with only a few general recommendations for approaching these studies suggested where appropriate or obvious. Specific recommendations for conducting most studies have not been made because we do not think that we are the best qualified to make them. Five of the recommended research topics deal with technical issues, the remaining three with social and economic issues. The most promising agricultural region in which to initiate studies is the Midwestern corn belt because it is by far the largest geographic region where this approach is appropriate (Figure 1) and it is the agricultural region with the poorest water quality (Omernik, 1977).

Recommended Studies

(a) Whole watershed demonstration studies are needed in several regions to establish the feasibility and utility of using restored wetlands as sinks for contaminants in agricultural runoff. The Food, Agriculture, Conservation and Trade Act of 1990 (S. 2830) established the Wetlands Reserve Prograin (section 1438) that allows the enrollment of up to 1,000,000 acres of land for restoring and protecting wetlands by 1995. This new program could be used with the agricultural water quality incentives (section 1439) of the 1990 Act to help fund the demonstration studies. Such watershed-level studies will provide essential information on (1) the costs of constructing wetlands; (2) the best way to establish wetlands in different kinds of agricultural landscapes; (3) how acceptable this approach is in different regions of the country; (4) how effective wetlands are as sinks for contaminants in agricultural runoff, and (5) how best to organize and administer a wetland restoration program. All the other recommended studies should be done as part of, or in conjunction with, these studies.

(b) Studies of the effectiveness of restored or recently created wetlands as sinks for nutrients and organic contaminants are essential. Such studies should determine the fate of different contaminants and their sustainable loading rates. These studies can most easily and economically be done using mesocosms, but field studies of the actual performance of restored or recently created wetlands also should be done. Reliable sustainable loading rates are essential for calculating the appropriate size of restored wetlands.

(c) Studies are also required of the effects of NPS contaminants on the development of restored wetlands and on existing wetlands. The most worrisome contaminants are sediments, herbicides, and insecticides. There are already many studies that suggest that contaminants can have adverse effects on wetlands (see Sheehan et al., 1987; Stuber, 1988; Facemire, no date). Dosage studies using mesocosms are the most logical approach to impact studies, and ideally impact studies should be combined with mesocosm loading studies. Field studies of impacts of contaminants on wetlands are also needed to examine spatial patterns of impacts within wetlands and long-term effects on wetlands.

(d) Landscape simulation models of the origin and movement of NPS pollutants in selected agricultural landscapes should be developed to determine the number of wetlands needed and where they should be located to reduce contaminants to acceptable levels. These models can be used to evaluate both environmental and economic effectiveness of different site selection criteria and wetland designs in diverse watersheds. If feasible, these models should be designed so that they can be used with minimal data for the routine planning of watershed-level restoration programs.

(e) Site selection and design criteria for created and restored wetlands need to be established. A literature review and synthesis plus expert opinions on siting and sustainable loadings from scientists in different regions of the country is the best approach to developing siting and design criteria. Among the questions that should be considered are: (1) in which regions of the country should this approach be used? (2) what are the major contaminants in agricultural runoff in each region? and (3) what will be the potential impacts of establishing wetlands on local and regional hydrology, particularly groundwater levels?

(f) Studies should be conducted of farmers' and local business and community leaders' attitudes toward a landscape approach to NPS problem reduction and targeting watersheds as units in government programs. This information is needed to develop educational programs for farmers on wetlands and water quality, to plan how best to implement landscape approach solutions, and to decide what public policy changes and administrative structures will be needed.

(g) Legal and public policy issues created by wetland restoration programs need to be researched. This is needed to develop effective interagency programs for administering, funding, and implementing these programs. One important legal issue that needs to be examined is the implication of disrupting drainage networks to establish wetlands. What are the liabilities for agencies and farmers who fund or restore wetlands if other landowners in the watershed believe that this will negatively affect drainage of their land? Disputes between adjacent landowners and government agencies over wetland restorations have already occurred in Iowa and resulted in the cancellation of restoration projects. Another important issue is whether restored or created wetlands that were established to treat agricultural runoff should be designated waters of the United States and thus become subject to Federal regulation.

(h) Studies are needed to determine the economic costs and benefits of this approach to rural NPS pollution reduction and how costs associated with it can be internalized. This information is critical if effective public policies are to be designed and implemented. Recreational, wildlife, and other ancillary values of having wetlands in agricultural landscapes should be considered in these economic analyses. These studies should be done in conjunction with both the modeling studies and the demonstration studies.

SUMMARY

The restoration of wetlands in appropriate agricultural landscapes as sinks for contaminants in runoff is feasible. Major technical issues that need to be resolved include the effects of contaminants, particularly sediments and pesticides, on wetland ecosystem composition, structure, and function; the fate of organic contaminants in wetlands; the development of site selection criteria; and the development of design criteria. Much of the work on nutrient and contaminant processing in wetlands has been done in natural or highly engineered wetland systems often designed for tertiary sewage treatment (Hammer, 1989). To be economically acceptable, wetlands in agricultural landscapes must be restored or created with minimal cost and effort. There is little work on organic contaminant processing in recently restored or created wetlands. There is even less work on the effect of contaminants and sediments on invertebrate production, denitrification rates, litter decomposition rates, species composition of vegetation, etc. There are many social, economic, and political barriers to the adoption of landscape-level approaches, such as wetland restoration, to solving NPS problems. These include having watersheds accepted as the basic landscape unit both legally and socially, establishing mechanisms at the landscape level to develop restoration programs and gain acceptance for them, and determining needed incentives for wetland restoration programs and ultimately their economic effectiveness, considering both private and public costs and benefits.

Eight research projects are identified that need to be completed before guidelines for establishing watershed-level wetland restoration programs can be developed. These include five technical projects (watershed-level demonstration projects, landscape or watershed simulation models, site selection criteria, sustainable loading rates for contaminants, and effects of contaminants on wetlands) and three economic and social projects (attitudes of farmers and rural leaders, legal and public-policy implications, economic costs and benefits).

ACKNOWLEDGMENTS

We would like to thank Roger Link of the Soil Conservation Service in Des Moines for information on various Federal and State programs dealing with water quality. We also would like to thank Bill Crumpton, Tom Jurik, Bob Kadlec, Vernon Meentemeyer, Roger Link, and Suzanne van der Valk for their comments on various drafts of the manuscript.

REFERENCES

Bowden, W. B., 1987. The biogeochemistry of nitrogen in freshwater wetlands. Biogeochemistry, 4: 313-348.

Braden, J. B., G. V. Johnson, A. Bouzaher, and D. Miltz, 1989. Optimal spatial management of agricultural pollution. American Journal of Agricultural Economics, 71: 404-413.

Dahl, T. E., 1990. Wetland losses in the United States 1780's to 1980's. U. S. Department of the Interior, Fish and Wildlife Service, Washington, DC.

Davis, C. B, J. L. Baker, A. G. van der Valk, and C. E. Beer, 1981. Prairie pothole marshes as traps for nitrogen and phosphorus in agricultural runoff. In: B. Richardson (ed.), Proceedings of the Midwest Conference on Wetland Values and Management. Freshwater Society, Navarre, MN.

Facemire, C. F., No date. Impact of agricultural chemicals on wetland habitats and associated biota with special reference to migratory birds: a selected and annotated bibliography. U.S. Fish and Wildlife Service and South Dakota Agriculture Experiment Station, South Dakota State University, Brookings, SD.

Grue, C. E., L. R. DeWeese, P. Mineau, G. A. Swanson, J. R. Foster, P. M. Arnold, J. N. Huckins, P. J. Sheehan, W. K. Marshall, and A.

P. Ludden, 1986. Potential impacts of agricultural chemicals on waterfowl and other wildlife inhabiting prairie wetlands: An evaluation of research needs and approaches. Transactions of the North American Wildlife and Natural Resources Conference, 51: 357-383.

Grue, C. E., M. W. Tome, G. A. Swanson, S. M. Borthwick, and L. R. DeWeese, 1988. Agricultural chemicals and the quality of prairie-pothole wetlands for adult and juvenile waterfowl-what are the concerns. pp. 55-64. In: P. J. Stuber (ed.), Proceedings of the National Symposium on Protection of Wetlands from Agricultural Impacts. U. S. Fish and Wildlife Service, Biol. Rep. 88(16), Washington, DC,

Hammer, D. A. (ed.), 1989. Constructed Wetlands for Wastewater Treatment. Lewis Publishers, Inc., Chelsea, MI.

Howard-Williams, C., 1985. Cycling and retention of nitrogen and phosphorus in wetlands: theoretical and applied perspective. Freshwater Biology, 15: 391-431.

Knisel, W. G., 1980. CREAMS: a field-scale model for chemical runoff, and erosion from agricultural management systems. Conservation Research Report No. 26. U. S. Department of Agriculture, Washington, DC.

Kusler, J. A. and M. E. Kentula (eds.), 1990. Wetland Creation and Restoration: The Status of the Science. Island Press, Washington, DC.

Lane, L. J. and M. A. Nearing, 1989. USDA-Water Erosion Prediction Project (WEPP). Hillslope Profile Model Documentation. NSERL Report No. 2. National Soil Erosion Research Laboratory, Agricultural Research Services, West Lafayette, IN.

Martin, D. B. and W. A. Hartman, 1987. Effect of cultivation on sediment composition and deposition in prairie pothole wetlands. Water Air and Soil Pollution, 34: 45-53.

Moody, D. W., 1990. Groundwater contamination in the United States. Journal of Soil and Water Conservation, 45: 170-179.

Neely, R. K. and J. L. Baker, 1989. Nitrogen and phosphorous dynamics and the fate of agricultural runoff. In: A. G. van der Valk (ed.), Northern Prairie Wetlands. Iowa State University Press, Ames, IA.

Novotny, V. and G. Chesters, 1989. Delivery of sediment and pollutants from nonpoint sources: a water quality perspective. Journal of Soil and Water Conservation, 44: 568-576.

Omernik, J. M., 1977. Nonpoint source-stream nutrient level relationships: A nationwide study. EPA-60013-77-105. U.S. Environmental Protection Agency, Corvallis, OR.

Pavelis, G. A., 1987. Economic survey of farm drainage. pp. 110-136. In: G. A. Pavelis (ed.), Farm Drainage in the United States: History, Status, and Prospects. Miscellaneous Publication No. 1455. Economic Research Service, U.S. Department of Agriculture.

Phillips, J. D., 1989. Fluvial sediment storage in wetlands. Water Resources Bulletin, 25: 867-873.

Ribaudo, M. O., D. Colacicco, A. Barbarika, and C. E. Young, 1989. The economic efficiency of voluntary soil conservation programs. Journal of Soil and Water Conservation, 44: 40-43.

Sheehan, P. J., A. Baril, P. Mineau, D. K. Smith, A. Harfenist, and W. K. Marshall, 1987. Impact of Pesticides on the Ecology of Prairie-Nesting Ducks. Technical Report Series 19. Canadian Wildlife Service, Environment Canada, Ottawa, Ontario, Canada.

Stuber, P. J., 1988. Proceedings of the National Symposium on Protection of Wetlands from Agricultural Impacts. U. S. Fish and Wildlife Service, Biol. Rep. 88(16), Washington, DC.

U.S. Environmental Protection Agency, 1990. Water Quality Standards for Wetlands: National Guidance. EPA 440/S-90-011. Office of Water Regulations and Standards, Environmental Protection Agency, Washington, DC.

van der Valk, A. G., S. D. Swanson, and R. F. Nuss, 1983. The response of plant species to burial in three types of Alaskan wetlands. Canadian Journal of Botany, 61: 1150-1164.

Young, R. A., C. A. Onstad, D. D. Bosch, and W. P. Anderson, 1987. AGNPS: a nonpoint source pollution model for evaluating agricultural watersheds. Journal of Soil and Water Conservation, 44: 168-173.

LIST OF FIGURES

Figure 1

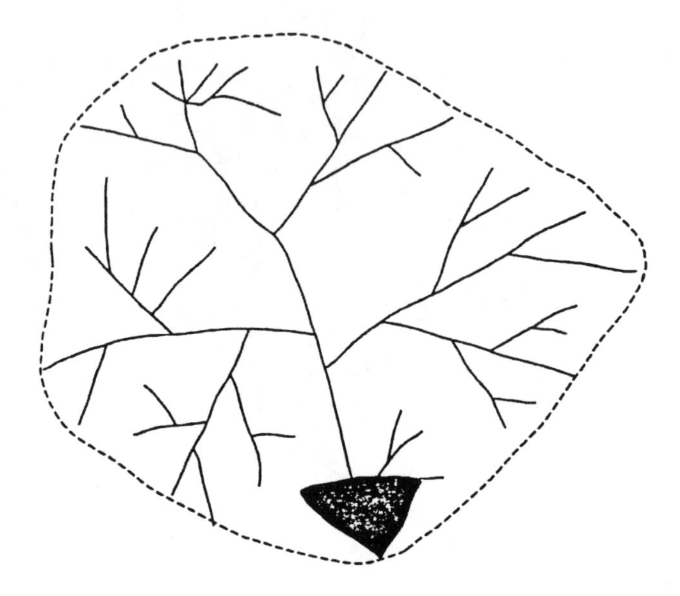

Figure 2

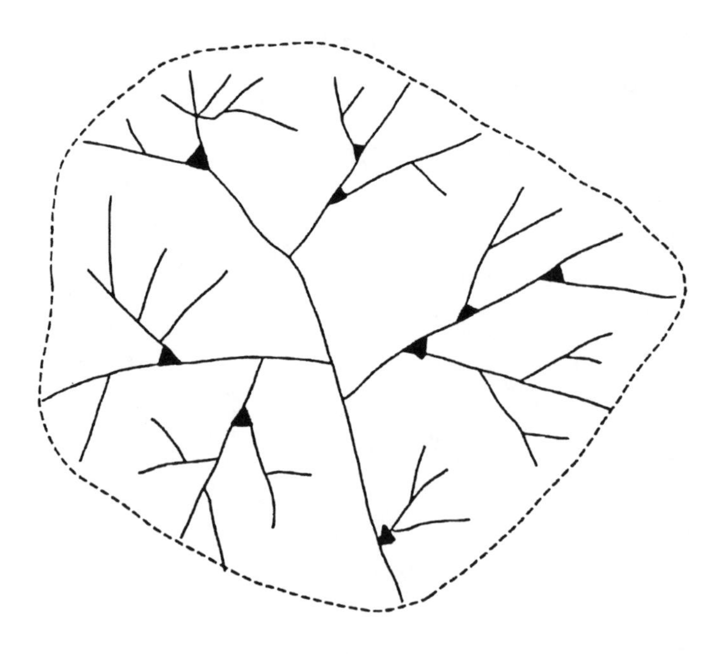

Figure 3

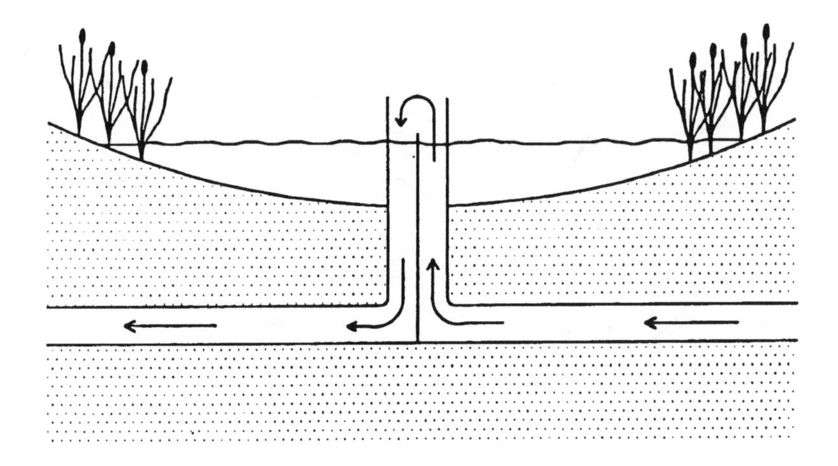

Figure 4

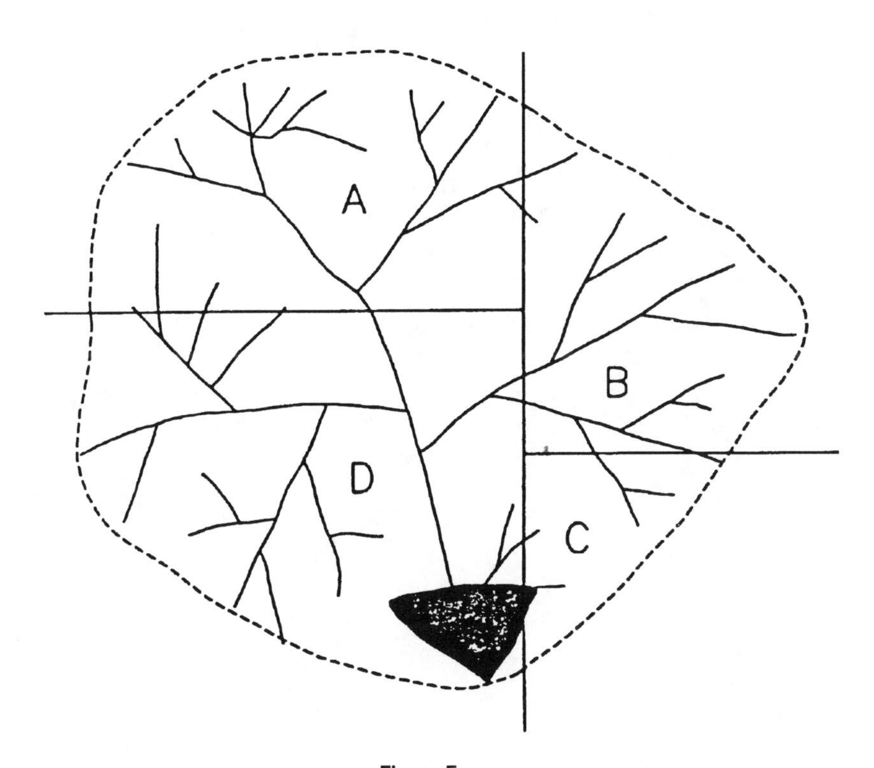

Figure 5

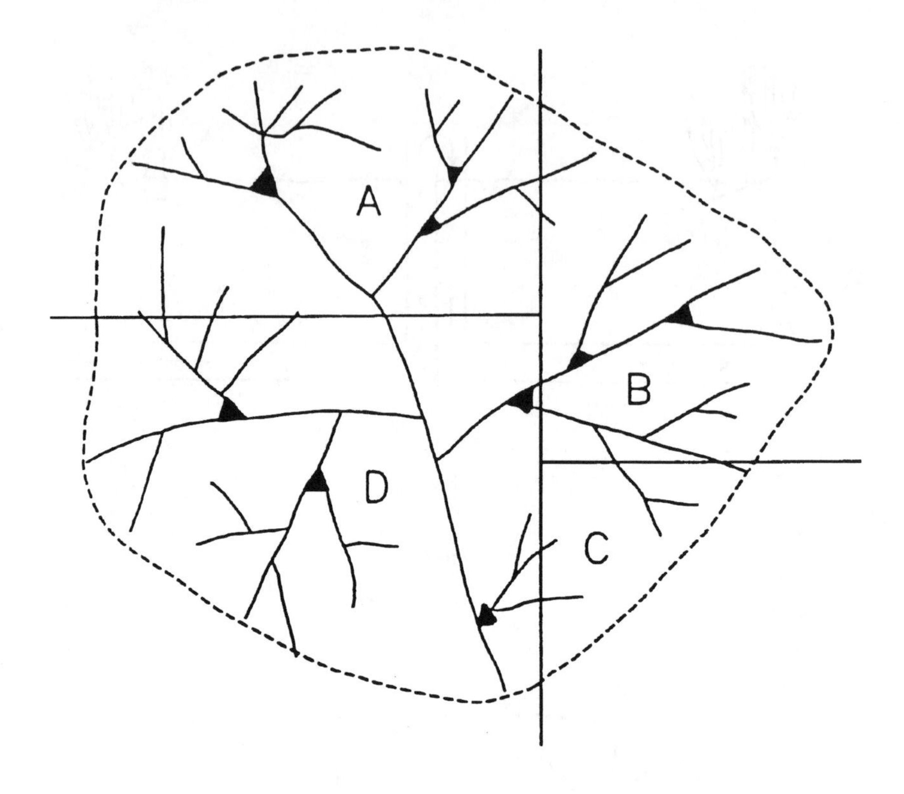

Figure 6

CHAPTER 10

Research and Information Needs Related to Nonpoint Source Pollution and Wetlands in the Watershed: An EPA Perspective

Beverly J. Ethridge (6E-FT), U.S. Environmental Protection Agency.
Richard K. Olson, ManTech Environmental Technology, Inc., U.S. EPA Environmental Research Laboratory

ABSTRACT

Two related Environmental Protection Agency (EPA) efforts, wetlands protection and nonpoint source pollution control, fail to fully consider landscape factors when making site-specific decisions. This paper discusses the relationship of the two programs and the use of created and natural wetlands to treat nonpoint source (NPS) pollution. Recommendations to improve the programs include increased technical transfer of existing information, and more research on construction methods and siting of created wetlands to effectively manage NPS pollution. Additional research is also needed to determine (1) the maximum pollutant loading rates to assure the biological integrity of wetlands, (2) the effectiveness of current land use practices in protecting habitat and water quality functions, (3) wetland functions as pollutant sinks, (4) NPS pollution threats to wildlife, (5) practical watershed models, and (6) indicators and reference sites for monitoring wetland condition. Model watershed demonstrations, jointly implemented by the research and conservation communities, are recommended as a means of integrating research results.

INTRODUCTION

The U.S. Environmental Protection Agency (EPA) oversees many regulatory efforts designed to protect U.S. surface waters, including wetlands (Fields, this volume). Some of these programs have overlapping objectives, and efforts are made to ensure that these programs are well coordinated during implementation. Coordination is achieved in part by emphasizing problem solving using a holistic approach; as is embodied in watershed-, landscape- or ecoregion-based initiatives.

A watershed approach is commonly used in programs to control nonpoint source (NPS) pollution. Through this approach, environmental agencies and individuals can efficiently target pollutant abatement activities within the watershed. Knowledge of spatial and functional relationships between wetlands and various land-use practices at the watershed scale helps to design integrated NPS control strategies.

The purpose of this paper is to identify information gaps that hinder the inclusion of wetlands, especially riparian wetlands, in NPS control strategies, and to recommend research and technology transfer actions to help fill these gaps.

BACKGROUND

Riparian areas in the southeastern United States are typically forested areas along stream sides that maintain the stability of the water course and buffer the parent waterbody from adjacent land uses. Riparian areas also provide significant habitat functions.

Riparian systems (riparian, as used in this paper, generally refers to eastern riparian systems, which are often wetlands) process and store large amounts of NPS pollutants from agricultural activities, particularly nutrients and sediment (Cooper et al., 1986; Fail et al., 1987; Lowrance et al., 1984a,b; Peterjohn and Correll, 1984; Schlosser and Karr, 1981; vhighaln et al., 1986). Lowrance et al. (1985) described the nutrient cycling processes of riparian areas as including (1) stabilizing sediments in stream banks, (2) long-term storage of nutrients in woody material, (3) uptake of nutrients from subsurface flow, and (4) denitrification. An earlier study illustrated the relationship between riparian ecosystems and water quality (Lowrance et al., 1983). In a Georgia coastal plain, total replacement of a riparian forest with a mix of crops similar to those grown on uplands was projected to result in an estimated twenty-fold increase in loading of NO_{3}-N to the stream.

While many studies of riparian wetlands and water quality have been conducted on individual sites or relatively small watersheds, there is much less information on the role of riparian areas at the landscape scale (Johnston et al., 1990; Schlosser and Karr, 1981; Kuenzler, 1989). Information on the impacts of NPS pollution on the ecosystem health and habitat functions of wetlands is also limited (van der Valk and Jolly, this volume).

Research to date has primarily addressed pollutant processing within existing riparian areas or pollutant fate within the landscape if riparian vegetation is removed. The extent to which structural restoration of riparian areas restores water quality functions is not well known (Kusler and Kentula, 1990). During the recent development of draft management measures for the newly reauthorized Coastal Zone Management Act, EPA staff did not gain management approval to include restoration of riparian systems as an enforceable technical measure to ensure water quality protection. This decision was based on the paucity of data in the literature on the water quality benefits of restored systems.

CLEAN WATER ACT

Much of EPA's legislative authority to regulate surface waters comes from the Clean Water Act. In accordance with the 1987 Amendments to the Clean Water Act (Section 319), activities resulting in significant NPS pollution must be identified and addressed through source reduction by application of best management practices (BMPs). In its 1989 *Nonpoint Sources: Agenda for the Future* (U.S. EPA, 1989), EPA indicated that agricultural activities account for more than half of all NPS impacts to surface waters, significantly impairing the quality and beneficial uses of the Nation's lakes, streams, rivers, and estuaries. These conclusions were based on the States' biennial water quality inventories, also know as Section 305(b) reports. Section 319 provides for funding State efforts to reduce water quality impacts. Sections 314 (clean lakes) and 320 (National Estuary Program) also place high priority on abating adverse effects from NPS pollution.

A particularly significant component of the Clean Water Act, Section 303, requires the establishment of total maximum pollutant loads within watersheds that will ensure protection and propagation of indigenous fish and wildlife resources. This section requires that all sources of pollution, including both point and nonpoint, be addressed.

Implementation of each of these Clean Water Act sections requires a watershed-based approach to problem solving. Even in the administration of Section 404 (dredge and fill permits), which has traditionally been applied to wetlands by EPA and the Corps of Engineers (COE) on a site-by-site basis, more emphasis is now being placed on cumulative impacts, or, more precisely, on the impacts to wetland systems within a watershed or landscape context.

The obvious relationship between wetlands and NPS issues led EPA to issue program guidance encouraging the linkage of NPS and wetland program objectives by joint implementation of projects (U.S. EPA, 1990). This has increased the need for technical guidance on the relationship between NPS pollution and wetlands in the many watershed improvement projects being initiated throughout the United States by the EPA, U.S. Department of Agriculture (USDA), Tennessee Valley Authority (TVA), COE, and State and local agencies (see Whitaker and Terrell, this volume).

An example of the need for coordination between NPS and wetlands programs is described by Kuenzler (1989). Small streams in a watershed usually comprise most of the total stream length. Their riparian systems, therefore, provide the watershed's longest cumulative lengths of buffer between the stream system and upland sources of NPS pollution. However, COE regulations allow deposition of fill material in isolated and headwater wetlands with minimal review, which places significant portions of these riparian wetlands at risk. The failure of wetland protection programs to adequately protect riparian wetlands along low-order streams makes it more difficult for NPS control programs to meet water quality objectives.

IDENTIFYING RESEARCH NEEDS

Regulatory agencies need additional information and research to effectively combine wetlands protection and NPS control strategies. As a step toward defining these needs, regional program managers and scientists from EPA as well as other Federal, State, and local agencies, and members of the research community were surveyed. Inquiries were sent to people currently involved in NPS program implementation, wetland protection program implementation, and related research.

Approximately 35 individuals responded to the survey, raising over 100 research and technology transfer issues, many of which were closely related and have been consolidated and summarized. Five main areas were identified where additional technical guidance and research is needed: (1) relationships between wetland water quality and habitat functions, (2) wetland water quality functions at the landscape scale, (3) monitoring and evaluation techniques, (4) design criteria for constructed wetland treatment systems, and (5) technology transfer. The most common research needs within each category are listed below.

Water Quality and Habitat Functions

- Determination of thresholds (i.e., pollutant loading rates and cumulative loadings) above which wetland structure and functions are degraded.
- Differences in loading thresholds for different wetland types and geographic regions.
- Long-term sustainability of wetland water quality functions.
- Recovery of water quality functions following perturbations such as severe flood events.
- Indirect effects of NPS pollution on the health of migratory waterfowl and wading bird populations (e.g., greater vulnerability to parasites and disease).

Landscape Functions

- Methods for determining pollutant loading rates to wetlands based on upland acreage, land uses, and spatial configuration.
- Landscape design: Should restoration of wetlands within the watershed attempt to duplicate historical wetland patterns?
- Site location criteria: Are wetlands near the pollutant source most critical, or are larger systems, strategically located-downstream, more important in the water quality relationship?
- Relationships between watershed hydrology, biogeochemistry, and wetland functions.
- Integration of wetlands with other BMPs for controlling NPS pollution.
- Models of landscape function for predicting water quality effects of different wetland protection and restoration scenarios.

Monitoring and Evaluation Techniques

- Indicators of wetland function and condition; needed at scales from site to regional, and for a variety of stressors including chemical, hydrological, and physical perturbations.
- Methods for monitoring the fate and effects of NPS pollutants within a watershed.
- Methods for selecting reference sites for use in monitoring programs.
- Procedures for evaluating the effectiveness of BMPs at improving water quality and protecting wetlands from degradation and NPS pollution.
- Post-restoration monitoring to help improve wetland restoration procedures.

Design Criteria for Constructed Wetland Treatment Systems

- Improved design and operational criteria for animal waste treatment systems.
- Improved design and operational criteria for field runoff treatment systems.
- Improved capabilities for predicting treatment system longevity and maintenance needs.

Technology Transfer

- Synthesis of research results from the literature, and transmittal to program managers in the form of clear and comprehensive guidance on integrating wetland and NPS regulatory programs.
- Additional watershed-level demonstrations of integrated NPS/wetland programs.
- Economic information relating the relative cost and benefits of different BMPs and wetlands protection strategies relative to water quality improvement.
- Improved collaboration and information exchange among Federal, State, and private groups.

SUMMARY

The range and number of research needs identified from the survey make clear the great need for research and technical guidance relating wetlands and NPS pollution. An opinion often expressed is that watershed-scale water quality improvement projects are being funded and implemented, yet these projects are proceeding without adequate guidelines from the Federal community. Program implementers must also work with farmers who are understandably reluctant to invest monies on efforts that lack clear, definable benefits. It is critical that farmers, other landowners, and researchers all be included as members of the committees that oversee research and development of technical guidance.

It is well documented that wetlands and riparian systems perform a significant role in protecting and maintaining water quality within the landscape. Considerable information is available showing that wetlands improve water quality by filtering and processing pollutants. However, that function has not been adequately described in holistic, watershed terms that also address other natural functions of wetlands, particularly the relationship between water quality functions and habitat quality.

These information gaps leave field program implementers (both NPS and wetland protection program implementers) with little guidance on how to implement wetland/riparian restoration efforts on a watershed scale. The gaps need to be filled by research, but just as importantly, the research results must be packaged and distributed in formats that can be easily used by land managers.

RECOMMENDATIONS

- EPA should take a more active role in technology transfer. Many of the "research" questions raised may, in fact, be answered to some extent in the literature, but the information is not available to program personnel in a useful form. An interagency team, including other Federal agencies, as well as State, local, and nongovernmental groups, should be established to determine the most effective means of synthesizing and disseminating information.
- Collaborative watershed demonstration projects between EPA and other groups should be implemented in multiple regions of the United States. Greater EPA participation in the USDA Hydrologic Unit Areas program and other USDA water quality improvement demonstration projects (see Whitaker and Terrell, this volume) is

one possible approach. These projects would provide a means of integrating a number of research tasks including:
- developing models that define the functional relationships between wetlands and other land uses in the watershed;
- developing models for predicting the effects of different BMPs and landuse patterns on the fate and transport of NPS pollutants;
- developing indicators (biological, chemical, etc.) of wetland condition, and identifying reference wetlands.

- USDA and TVA should continue their leadership roles in developing improved design criteria for agricultural wastewater constructed wetlands. EPA should assist by assessing the ecological condition and effects of these systems.

These recommendations are steps toward filling some of the information gaps that are hindering the effective coordination of NPS control and wetlands protection programs. A strong link between science and policy is necessary if the multiple objectives of these two programs are to be met.

REFERENCES

Cooper, J.R., J.W. Gilliam, and T.C. Jacobs, 1986. Riparian areas as control of nonpoint pollutants. pp. 166-192. In: D.L. Correll (ed.), Watershed Research Perspectives. Smithsonian Institution Press, Washington, DC.

Fail, J.L., Jr., B.L. Haines, and R.L. Todd, 1987. Riparian forest communities and their role in nutrient conservation in an agricultural watershed. American Journal of Alternative Agriculture, 2(3): 114-121.

Johnston, C.A., N.E. Detenbeck, and G.J. Niemi, 1990. The cumulative effects of wetlands on stream water quality and quantity. A landscape approach. Biogeochemistry, 10: 105-141.

Kuenzler, Edward J., 1989. Value of forested wetlands as filters for sediments and nutrients. pp. 85-96. In: Proceedings of Symposium on the Forested Wetlands of the Southern U.S., Orlando, FL.

Kusler, Jon A. and Mary E. Kentula (eds.), 1990. Wetland Creation and Restoration: The Status of the Science. Island Press, Washington, DC. 591 pp.

Lowrance, R.R., R.L. Todd, and L.E. Asmussen, 1983. Waterborne nutrient budgets for the riparian zone of an agricultural watershed, Agriculture, Ecosystems and Environment, 10: 371-384.

Lowrance, R.R., R.L. Todd, and L.E. Asmussen, 1984a. Nutrient cycling in an agricultural watershed: II. Streamflow and artificial drainage. Journal of Environmental Quality, 13(1): 27-32.

Lowrance, R., R. Todd, J. Fail, Jr., O. Hendrickson, Jr., R. Leonard, and L. Asmussen, 1984b. Riparian forests as nutrient filters in agricultural watersheds. BioScience, 34(6): 374-377.

Lowrance, R., R. Leonard, and J. Sheridan, 1985. Managing riparian ecosystems to control nonpoint pollution. Journal of Soil and Water Conservation, 40(1): 87-91.

Peterjohn, W.T. and D.L. Correll, 1984. Nutrient dynamics in an agricultural watershed: Observations on the role of a riparian forest. Ecology, 65(5): 1466-1475.

Schlosser, I.J. and J.R. Karr, 1981. Water quality in agricultural watersheds: Impact of riparian vegetation during base flow. Water Resources Bulletin, April: 233-240.

U.S. Environmental Protection Agency, 1989. Nonpoint Sources: Agenda for the Future. WH-556, U.S. EPA, Office of Water, Washington, DC. 31 pp.

U.S. Environmental Protection Agency, 1990. National Guidance: Wetlands and Nonpoint Source Control Programs. U.S. EPA, Office of Water Regulations and Standards and Office of Wetlands Protection, Washington, DC.

Whigham, D.F., C. Chitterling, B. Palmer, and J. O'Neill, 1986. Modification of runoff from upland watersheds—the Influence of a diverse riparian ecosystem. pp. 305-332. In: D.L. Correll (ed.), Watershed Research Perspectives. Smithsonian Institution Press, Washington DC.

CHAPTER 11

Federal Programs for Wetland Restoration and Use of Wetlands for Nonpoint Source Pollution Control

Gene Whitaker, U.S. Fish and Wildlife Service, Division of Habitat Conservation.
Charles R. Terrell, Ecological Sciences Division, Soil Conservation Service.

ABSTRACT

A review of Federal wetlands programs shows that a number of agencies have significant wetland restoration and creation efforts. Water quality improvement is not the main objective of most of these programs, and areas with high nonpoint source (NPS) pollution may actually be avoided in order to protect wetland values such as habitat. However, ancillary water quality benefits are provided by many created and restored wetlands, and agencies such as the U.S. Department of Agriculture are actively evaluating the use of created and restored wetlands as components of NPS control strategies.

INTRODUCTION

Wetland restoration has come into widespread use in the last few years. All Federal land management agencies now have active programs to restore wetlands on lands under their control, and most of them have programs to assist other agencies and private landowners restore wetlands. In addition, several U.S. Department of Agriculture (USDA) agencies and the Department of the Interior (DOI) Fish and Wildlife Service (FWS) are implementing specifically mandated programs to help private landowners enhance, restore, and create wetlands. All of these agency programs contribute to the control of nonpoint source (NPS) pollution. However, control of NPS pollution is only one objective of most of these programs, and, in most cases, it is not a major objective. This paper briefly reviews the goals and objectives of the restoration programs of the major Federal agencies and what they are doing (see U.S. EPA (1989) for additional details).

In all these agency programs, the definition between wetland restoration and wetland enhancement is blurred. Restoration can be anything from blocking a drainage ditch that only removed a small fraction of the water from a wetland to major efforts that restore wetlands drained many years ago. The objective of most restorations is to recover the original wetland type, size, value, and vegetative community that existed before man's activities. Wetland enhancement refers only to the restoration of partially damaged wetlands. However, enhancement may be improving the value of a wetland for a function considered by the planners to be more important. The following definitions of these terms recently adopted by the Soil Conservation Service (SCS) are recommended for use (USDA, 1991a):

> *Wetland restoration* is defined as the rehabilitation of a degraded existing wetland or a hydric soil area that was previously a wetland.
>
> *Wetland enhancement* is defined as the improvement, maintenance, and management of existing wetlands for a particular purpose or function, often at the expense of others.
>
> *Wetland creation* is defined as the conversion of a non-wetland area into a wetland where a wetland never existed.
>
> *Constructed wetlands* are specifically designed to treat both nonpoint and point sources of water pollution.

SUMMARY OF FEDERAL WETLAND PROGRAMS

Federal Land Management Agencies

Eight Federal agencies manage most of the wetlands and sites for wetland restorations owned by the Federal government. Each has its own programs for restoring and enhancing wetlands. Most importantly, each has its own legislatively mandated missions and wetland goals and objectives. All the land management agencies emphasize the development of land management plans, including wetland restorations, with a high level of local public input. They are responsive to the needs and desires of the local communities,

Department of Defense

The Army, Navy, Air Force, and Coast Guard control approximately 25 million acres containing many acres of wetlands and sites where wetlands could be restored. As compatible with their primary mission, the land management plans for each installation address wetland management and may include wetland restorations in cooperation with other State and Federal agencies. On many bases, wetlands are restored in cooperation with the Fish and Wildlife Service to improve habitat for migratory birds. On many bases, the restoration and conservation of rare and unique wetland ecosystems is emphasized.

U.S. Army Corps of Engineers

The Corps manages 11 million acres of Federal land. Management of these lands, including wetlands and wetland restorations, ". . . is directed toward the continued enjoyment and maximum sustained use of public lands, waters, forests, and associated recreational resources consistent with their aesthetic and biological values . . ." Most of the Corps' wetland restorations are aimed at replacing the wetland functions and values lost during the construction of major projects. They also have an active wetland restoration research program and provide training to other agencies on wetland restoration.

USDA Forest Service

The estimated 9 million acres of wetlands on the 190 million acres of national forest land are managed and restored for their multiple-use values. They have a very active program to restore riparian wetlands providing significant control of NPS pollution.

DOI Bureau of Reclamation

Wetlands are managed and restored on the 8.5 million acres of land managed by the Bureau of Reclamation (BOR) as an integral part of total water resources management. Wetland conservation focuses on ". . . fish and wildlife habitat with equal consideration given to functions such as sediment control, water and wastewater treatment, flood storage, and ground water recharge." Most of BOR's wetland restoration activity is aimed at replacing wetland functions and values lost as a result of their irrigation water delivery systems. BOR staff are actively researching methods to minimize the effects of irrigation return water on wetlands and ways to minimize adverse impacts. They are also researching wetland construction techniques for various purposes and are preparing a wetland manual with scientific and engineering guidelines and procedures related to wetlands restoration and construction.

DOI National Park Service

Wetlands in units of the national park system are managed, protected, and restored to maintain their original natural characteristics to the fullest extent possible. To accomplish this, the National Park Service has an active water resource program designed to provide adequate water quantity and water quality for protecting park wetlands. The quality of water entering the parks is their single greatest problem.

DOI Bureau of Land Management

The Bureau of Land Management (BLM) manages 178 million acres of land in the lower 48 states containing about 1 million acres of marshes, ponds, reservoirs, and lakes and 41,000 miles of streamside riparian wetlands. BLM has an active program to restore and manage wetlands and riparian areas on their lands, mainly in the arid west. BLM states that "The objective of riparian area management is to maintain, restore, or improve riparian values to achieve a healthy and productive ecological condition for maximum long-term benefits." A major objective of many of their riparian area management plans is to control NPS pollution by reducing the delivery of sediment to downstream reservoirs.

DOI Fish and Wildlife Service

The National Wildlife Refuge System has over 90 million acres of land in 462 refuges. The largest portion is in Alaska. However, over 25 million acres are in the lower 48 states. Most of the wetlands on refuges in the lower 48 states have been restored or enhanced during the last 20 years, primarily to improve habitat for waterfowl and other migratory birds. Active programs are underway in the major waterfowl breeding, migratory, and wintering areas to acquire and restore wetlands. Although control of NPS pollution is not a major purpose of wetland restorations on refuges, many refuge wetlands are adversely impacted by NPS pollution.

Sediment in incoming water is shortening the effective life of many refuge wetlands, and contaminants, especially in irrigated areas, are making some refuge wetlands unusable.

Federal Agencies with Private Lands Programs

Several Federal agencies have programs targeted to protect, enhance, and restore wetlands on private lands. The following sections list the Federal agencies with major programs and a brief discussion of each program and its goals and objectives (see U.S. DOI (1991) for additional details on the programs of Interior bureaus). The U.S. Environmental Protection Agency (EPA) has several programs, mainly working through State agencies, to restore wetlands. However, these are not discussed in this paper.

DOI Fish and Wildlife Service

The Fish and Wildlife Service (FWS), through its Private Lands Program, uses numerous avenues to effect the restoration and protection of wetlands on private lands throughout the country. Although restoration of wildlife habitat, especially for migratory birds, is a major objective in all the activities, emphasis is placed on the restoration and protection of the wide array of wetland functional values. However, the primary purpose of the restorations is to provide the maximum benefits for the longest possible time to wildlife and the people that enjoy them. Restoration sites are avoided where there is significant NPS pollution or other contaminant sources in the watershed, or measures are taken to avoid adverse impacts. Over the last 3 years, a total of more than 90,000 acres of mostly small wetlands have been restored, using funds appropriated for this purpose.

A concentrated effort is being made under the North American Waterfowl Management Plan (NAWMP) to protect a minimum of 2 million acres of existing wetland habitat in the United States by the year 2000. Under the plan, several million additional acres of wetlands will be restored and enhanced for migratory birds in Canada, Mexico, and the United States. The primary goal of the NAWMP is "To enhance and protect high-quality wetland habitat in North America that supports a variety of wetland-dependent wildlife and recreational uses." The NAWMP is being implemented through innovative Federal-State-private partnerships within and between States and Provinces.

DOI Bureau of Mines

The Bureau of Mines actively supports the construction of wetlands to help treat acidic mine water and regularly presents workshops on wetland construction for acid mine drainage treatment. They have developed design criteria for sizing wetlands based on the volume of acid mine drainage.

DOI Office of Surface Mining

The Office of Surface Mining strongly encourages and provides guidance on the development of wetlands for the broad range of natural functional values as part of all surface mine reclamations. They also have an active constructed wetlands research program to develop better recommendations.

USDA Cooperative Extension Service

The national office provides leadership, encouraging each State extension service to promote and provide assistance to private landowners for the management and restoration of wetlands. Most State extension services do have active programs promoting wetland conservation and provide information and technical assistance to landowners.

USDA Agricultural Stabilization and Conservation Service

Under the Cropland Reserve Program (CRP) of USDA's Agricultural Stabilization and Conservation Service (ASCS), approximately 950,000 acres of cropped wetlands and associated uplands have been re-established in natural vegetation under 10-year contracts. In addition, a couple million acres of natural and partially drained wetlands that were cropped when dry enough are included in fields entered in the CRP under the highly erodible criteria, generally because of wind erosion. More than 13,000 acres of wetlands have been restored under the CRP, in addition to wetlands the FWS has worked with farmers to restore on lands in the program, mainly for their wildlife and water quality benefits.

The Waterbank Program administered by the ASCS is presently protecting, in agricultural areas, 480,000 acres of natural wetlands and adjacent buffer areas under 10-year rental agreements.

The Wetland Reserve Program (WRP), authorized by the 1990 Farm Bill, is scheduled for implementation starting early, in 1991, pending authorization of funds. The objective of the WRP is to restore and protect, through easements, up to 1 million acres of wetlands in cropland on the Nation's farms and ranches. The law requires that priority shall be placed ". . . on acquiring easements based on the value of the easement for protecting and enhancing habitat for migratory birds and other wildlife." Technical assistance for the WRP is being provided by the FWS and the SCS. The emphasis will be on restoring wetlands to natural conditions for the multiplicity of functions and values provided. Areas with a high level of NPS pollution will be avoided to prolong the useful life of the wetlands and protect wildlife and other functional values of the restored wetlands.

Under the Agricultural Conservation Program (ACP), ASCS will cost share with farmers up to 75 percent of the cost of numerous practices that help control NPS pollution. For the "Creation of Shallow Water Areas" (wetland restoration) practice, cost share has been provided for the restoration of 555,000 acres of wetlands over the last 30 years. A new practice, eligible for up to 75 percent cost share and titled "Constructed Wetlands for Agricultural Waste Water Treatment (WP6)," is being designed to encourage farmers to use artificially created wetlands to control pollution by animal wastes. These constructed wetlands will be specifically designed as waste treatment systems that possess wetland characteristics. Their primary purpose will be to control NPS pollution. However, they will have some wetland wildlife value and provide some of the other functional values of natural wetlands. The technical requirements for their design will be supplied by the SCS, and SCS specialists will do the actual designs and supervise construction. Monitoring will be required for each constructed wetland installed.

USDA Soil Conservation Service

The SCS, with offices covering every county in the country, provides technical assistance to private landowners for wetland restoration. SCS provides detailed training to their field personnel in the planning and design of wetland restorations. SCS also helps develop the plans for some of the FWS's restorations and helps train FWS field people in the engineering aspects of designing wetland restorations. Working cooperatively with the FWS, Corps of Engineers, and EPA, SCS has just completed a detailed manual for the restoration, enhancement, and creation of wetlands (USDA, 1991a).

Interagency USDA Activities to Control NPS Pollution

The SCS with the Extension Service and the ASCS has been preparing to provide increased technical assistance for agricultural water quality problems. In February 1990, SCS issued its "Water Quality and Quantity Five-year Plan of Operations" (USDA, 1990a), which was followed water that year by the USDA "Policy for Water Quality Protection (USDA, 1990b)."

Hydrologic Unit Areas (HUAs)

In 1990, 37 HUAs were selected across the country where farmers and ranchers can participate in correcting water quality problems in their agricultural operations. In 1991, 37 additional HUAs were added to the list for a total of 74. Hydrologic Unit Areas are located in agricultural watersheds where the goal is to provide increased assistance to farmers and ranchers in voluntarily applying conservation practices to improve the water quality of the area (USDA, 1991b).

In each area, cost-sharing provides agricultural operators with the incentive to install conservation practices, such as animal waste control facilities, grassed waterways, irrigation water management systems, or integrated crop management for water quality improvement. Costshare funds are from the ASCS and State programs. SCS provides technical assistance to farmers and ranchers in the HUAs and the Extension Service provides information and educational assistance, including specific recommendations on the use of nutrients and pesticides (USDA, 1990c).

Demonstration Projects

In 1990, eight demonstration projects were approved across the country with an additional eight added in 1991. These projects demonstrate new ways to minimize the effects of agricultural NPS pollution, including the effects of nutrients and pesticides on groundwater. The goals are to demonstrate cost-effective water quality-oriented practices that can be used and shared by farmers and ranchers. Also, the projects are to accelerate the adoption of newly developed water quality technology. These projects are implemented under the joint leadership of SCS and the Extension Service, with financial assistance provided by ASCS.

ACP Water Quality Special Projects

In this program, funds are reserved by ASCS at the national level to fund Water Quality Special Projects developed by county Agricultural Stabilization and Conservation Committees. Project emphasis is on improving the quality of surface and ground waters that are impaired by agricultural NPS pollution. The projects are administered by ASCS with educational and technical assistance from the Extension Service and SCS.

It is obvious that overall integration between SCS, ASCS, and the Extension Service is necessary to accomplish the needed goals of agricultural water quality improvement. To that end, flexibility is built into the above three project types. For example, ACP Special Water Quality Projects can be used to solve water quality problems identified in HUAs and demonstration projects and also those identified locally, that may provide significant public benefits to nonagricultural interests. These projects may be designed to support State NPS objectives developed to meet the requirements of Section 319 of the Water Quality Act (see Fields, this volume).

SCS CONSTRUCTED WETLANDS PROGRAMS

The term "constructed wetland" is one that has come into use in recent years. Constructed wetlands, as we use the term, are constructed on non-wetland sites for the specific purpose of water quality improvement.

This definition distinguishes constructed wetlands from restored wetlands, which are "restored" or "recreated" where they once had existed. Created wetlands generally refer to wetlands developed where none previously existed to provide a variety of functional values.

This definition of constructed wetlands does not distinguish between point and nonpoint source pollution treatment. Constructed wetlands are used for treating both point and nonpoint sources of pollution. To a constructed wetland, the source of pollution does not make much difference. The wetland will do the job of cleaning water no matter what the source. Also, the definition does not mention surface and ground waters for similar reasons; water, whether above or below ground, really is one resource.

Basic Design Considerations

This paper will touch on this subject briefly because other papers in this volume (Hammer, Mitsch) and Cooper and Firldlater (1990) cover the design and construction of constructed wetlands in detail. There are four basic considerations relative to a constructed wetland and its functioning: (1) Construction, (2) Water, (3) Plants, and (4) Water Quality Improvement. Soils can be part of the equation or they can be left out entirely. For example, a constructed wetland can be lined with a synthetic liner or built in a concrete trough. Water hyacinths with their floating aspect never need to touch the soil.

Construction design and water considerations are key to a successful constructed wetland. Constructed wetland design is tied to water height; it is difficult to separate the two components. Water height, water residence time in the wetland, and exit of water from the wetland are critically important engineering considerations. The specifics of these characteristics must be developed if the constructed wetland is to have the desired functions.

Plants, the third component, need some examination. A misconception is that constructed wetlands in cold climates do not work. However, the evidence does not confirm that viewpoint. While it is true that photosynthesis and other plant functions are reduced in cold climates, wetland plants even below the snow or ice keep working, if at a reduced level. Actually, it appears that snow and ice act as an insulating blanket, so vital plant functions continue. Sunlight can penetrate ice to allow some photosynthesis. Further, it should be remembered that, in agricultural situations, water quality cleanup is needed largely in spring and summer and into autumn when organic production is at its peak and wetland plants are also very active.

Finally, the type of water quality improvement needed is vitally important and must be thoroughly understood before a shovel of soil is moved. The major decision to be made is whether the constructed wetland is for protection or a cure. Will the constructed wetland act in concert with other conservation practices to achieve an overall water quality improvement, or will the wetland serve as a device of last resort when other conservation practices are unable to achieve the desired goals of water quality improvement?

Landscape Considerations

It is possible to design for the four characteristics discussed above; but additionally a constructed wetland should fit the landscape. Does a constructed wetland need to be a perfect square or a rectangle? Consider the shapes of wetlands that Nature has provided. Is the constructed wetland to be located in the Midwest where ox-bows from old river bends are common? Why not create the new wetland in a similar shape? Is the wetland going to be located in Appalachia, where drowned river valleys are long and narrow? How about that shape? Rather than being bound by tradition that may dictate a so-called "standard" shape for a constructed wetland, today's designers should consider how to make the wetland an integral part of the land, not just another geometric configuration.

Regarding siting, some believe that constructed wetlands can be located at the bottom of the farmer's land, just before field runoff water enters the stream, river, or pond. Basically, this makes the wetland a structure of last resort and may lessen the perceived need to address water quality on a field-by-field basis. Without other measures, the constructed wetland would offer the only chance for cleanup before the runoff merges with receiving waters. It offers the tempting possibility to the farmer or rancher that other conservation practices need not be employed: "The constructed wetland will take care of my nonpoint source problems," he says.

An analogy can be made with a sewerage treatment plant. What happens with human waste is that clean water is intentionally contaminated with human waste and sent through a pipe; the contaminated water is intercepted at a sewerage treatment plant, often with the intention of cleaning the water just before that water empties into receiving waters. Thus, the sewerage treatment plant is a device of last resort. While alternative methods could be used that would not contaminate the water in the first place, they often are not. In some cases, millions of additional dollars are spent to protect the receiving waters because those same waters are used by the next town downstream for drinking purposes. Therefore, we must be absolutely certain that the sewerage treatment plant works 100 percent of the time or we endanger ourselves and our neighbors because of contaminated water.

As with the treatment plant, a constructed wetland located at the bottom of the farmer's land or at the bottom of the watershed becomes a structure of last resort. The wetland must work 100 percent of the time in all kinds of weather and under all conditions. This is an overly optimistic expectation. As with soil erosion protection, water quality protection must be implemented on a field-by-field basis. This practice reduces the pollution load at the edges of fields and does not transfer an increasingly larger load to the bottom of the watershed where one constructed wetland might be the only means of purifying the water before it empties into receiving waters.

Ongoing Activities

Presently, the SCS Plant Material Centers are conducting activities that enhance or restore existing wetlands or help establish constructed wetlands. These activities include: (1) developing propagation, establishment, and management techniques for potentially useful native wetland plants; (2) developing commercial sources of native wetland plants; (3) identifying the best plants and methodologies for using wetland plants for improving water quality associated with agricultural operations; (4) using wetland plants for drawdown areas of large impoundments; and (5) restoring coastal marshes and wetland riparian areas. These activities are planned or being conducted in Mississippi, Georgia, New Jersey, Michigan, Kansas, Oregon, Idaho, New York, and Texas. Most of this work involves cooperative projects with other Federal and State agencies.

In Mississippi and Alabama, SCS is working with the Tennessee Valley Authority and other agencies to see how constructed wetlands can be used to treat animal waste problems associated with livestock and poultry (Hammer, in this volume, and Hammer, 1989). Generally, in these types of situations the constructed wetland is used in combination with other structures, such as anaerobic and aerobic lagoons. The wetland acts as a polishing filter because the wetland would not be able to receive animal wastes directly.

In Maine, seven "Nutrient/Sediment Control Systems" have been constructed. Each system combines a series of conservation practices, including a constructed wetland that minimizes NPS pollution leaving the farms. The objective is to intercept cropland runoff from potato fields in the northernmost county of the lower 48 states, treat that runoff, and release clean water to a lake or stream. The basic system consists of a sediment basin, level-lip spreader, primary grass filter, constructed wetland, pond, polishing grass filter, and outlet (see Hammer, this volume, for details). Typical results show total phosphorus and suspended solids reduced by more than 90 percent, even during intense storm events (Wengrzynek and Terrell, 1990).

In Maryland, SCS is working with the State Department of Natural Resources on using constructed wetlands to control urban NPS runoff problems. These shallow marshes are created to remove pollutants, increase wildlife habitat, and provide educational and recreational opportunities. Maryland has designed and built more than 50 of these marshes across the state. Monitoring data show that total suspended solids were reduced by 63 percent; nitrate, 40 percent; total nitrogen, 12 percent; and total phosphorus, 38 percent.

In Pennsylvania, SCS has been working with the DOI Office of Surface Mining and the Bureau of Mines in the construction of new wetlands on abandoned coal mine lands. Several new wetlands have been constructed in the Pittsburgh-Sumerset area. These constructed wetlands were built on non-wetland soils to improve the quality of the water coming from abandoned coal mines. Monitoring data show that 90 percent of some pollutants such as iron are removed. The Bureau of Mines, which is the research arm of the 1977 Abandoned Mine Lands Act, maintains monitoring stations on these lands. Similar activity is ongoing in Kentucky and West Virginia.

CONCLUSIONS

In conclusion, we would like to point out several cautions and considerations to researchers and those charged with developing and implementing programs using constructed wetlands to control NPS pollution.

Single-Purpose vs. Multiple-Purpose Wetlands

Wetlands serve many valuable purposes to society; assimilation of various pollutants is just one of them. As this paper has discussed, most Federal agencies' wetland management and restoration programs are concerned about wildlife benefits of wetlands and the multiplicity of other functional values provided. They are concerned about the long-term maintenance of healthy functioning wetland ecosystems. Assimilation of moderate amounts of nutrients, pesticides, and other contaminants is a normal function of wetlands. However, in most places where there is a recognizably serious NPS problem, the water going into or through a wetland may reduce the other functional values of the wetland and shorten its useful life.

On the other hand, it is generally recognized that wetlands constructed specifically for the purpose of controlling NPS pollution do provide other benefits to society. SCS has often heard the comment, "You are constructing wetlands under the guise of water quality improvement so there will be more wetlands; water quality is only a ruse to get more wetlands for wildlife." While more benefits than water quality improvement can be realized from building constructed wetlands, such as wildlife habitat enhancement, there also are other advantages to constructed wetlands that have little to do with wildlife, such as flood control and recreation. One of the farmers collaborating with SCS was glad to have the water quality of the runoff leaving his fields improved, but he was even more delighted that he could raise baitfish in his constructed wetland with which to go fishing. Different people appreciate the different functions of wetlands and want constructed wetlands for different reasons. We should be open to those varied reasons, respect them, and see how constructed wetlands can accommodate a variety of needs, functions, and uses without impacting the functions and values of natural wetlands.

Section 404 Considerations

Farmers will not invest in constructed wetlands if they fear that in a few years a new generation of thinking might say that it looks like a wetland so it must be a wetland and therefore it falls under 404 (see Fields, this volume, for a description of regulations under Section 404 of the Clean Water Act). The wetland/environmental community must be able to assure farmers and ranchers who honestly invest in a constructed wetland as a conservation practice that they will not find themselves facing Section 404-related problems in the future.

The Future

Constructed wetlands offer an excellent potential for agriculturally related water quality improvement, but we do not know all the answers as to how they work and how they can work most effectively in agricultural situations. We need to foster more research and investigations that will provide some of these answers, while at the same time allowing us to proceed with using constructed wetlands as one strategy for controlling NPS pollution. The investigations must be conducted not only on a scientific basis, but also on a policy basis. Even though we do not have all of the answers to questions concerning constructed wetlands, we should continue to think of them as one more tool in the toolbox of conservation practices, rather than as a panacea that will cure all of our water quality problems.

REFERENCES

Cooper, P. F. and B. C. Findlater (eds.), 1990. Constructed Wetlands in Water Pollution Control: Proceedings of the International Conference on the Use of Constructed Wetlands in Water Pollution Control, Cambridge, England, Sept. 24-28, 1990. Pergamon Press.

Hammer, D. A. (ed.), 1989. Constructed Wetlands for Wastewater Treatment: Municipal, Industrial and Agricultural: Proceedings of the First International Conference on Constructed Wetlands for Wastewater Treatment, Chattanooga, TN, June 13-17, 1988. Lewis Publishers, Inc., Chelsea, MI.

U.S. Department of Agriculture, 1990a. Water Quality and Quantity: SCS Five-Year Plan of Operations, October 1, 1989-September 30, 1994. SCS National Bulletin No. 460-0-1.

U.S. Department of Agriculture, 1990b. USDA Policy for Water Quality Protection. Soil Conservation Service (SCS) National Bulletin No. 460-1-1.

U.S. Department of Agriculture, 1990c. Water Quality Education and Technical Assistance Plan: 1990 Update. Agricultural Information Bulletin No. 598.

U.S. Department of Agriculture, 1991a. Chapter 13. Wetland Restoration, Enhancement, or Creation. SCS Engineering Field Handbook. In press. 94 pp.

U.S. Department of Agriculture, 1991b. Water Quality Activities of the Soil Conservation Service: 1991 Update.

U.S. Department of the Interior, 1991. Wetland Stewardship; Highlights of the Department of the Interior's 1990 Wetlands Activities. Department of the Interior, Washington, DC. 37 pp.

U.S. Environmental Protection Agency, 1989. Wise Use and Protection of Federally Managed Wetlands: The Federal Land Management Agency Role. Results of a Workshop. U.S. Environmental Protection Agency, Washington, DC. 120 pp.

Wengrzynek, R. J. and C. R. Terrell, 1990. Using constructed wetlands to control agricultural nonpoint source pollution. In: P. F. Cooper and B. C. Findlater (eds.), Constructed Wetlands in Water Pollution Control: Proceedings of the International Conference on the Use of Constructed Wetlands in Water Pollution Control, Cambridge, England, Sept. 24-28, 1990. Pergamon Press.